T0325155

CONSIDERING ANIMALS

Considering Animals draws on the expertise of scholars trained in the biological sciences, humanities, and social sciences to investigate the complex and contradictory relationships humans have with nonhuman animals. Taking their cue from the specific "animal moments" that punctuate these interactions, the essays engage with contemporary issues and debates central to human-animal studies: the representation of animals, the practical and ethical issues inseparable from human interactions with other species, and, perhaps most challengingly, the compelling evidence that animals are themselves considering beings. Case studies focus on issues such as animal emotion and human "sentimentality"; the representation of animals in contemporary art and in recent films such as *March of the Penguins*, *Happy Feet*, and *Grizzly Man*; animals' experiences in catastrophic events such as Hurricane Katrina and the SARS outbreak; and the danger of overvaluing the role humans play in the earth's ecosystems. From Marc Bekoff's moving preface through to the last essay, *Considering Animals* foregrounds the frequent, sometimes uncanny, exchanges with other species that disturb our self-contained existences and bring into focus our troubled relationships with them. Written in an accessible and jargon-free style, this collection demonstrates that, in the face of species extinction and environmental destruction, the roles and fates of animals are too important to be left to any one academic discipline.

Considering Animals
Contemporary Studies
in Human–Animal Relations

Edited by

CAROL FREEMAN,
University of Tasmania, Australia

ELIZABETH LEANE,
University of Tasmania, Australia

and

YVETTE WATT
University of Tasmania, Australia

Routledge
Taylor & Francis Group

LONDON AND NEW YORK

First published 2011 by Ashgate Publishing

Published 2016 by Routledge
2 Park Square, Milton Park, Abingdon, Oxon OX14 4RN
711 Third Avenue, New York, NY 10017, USA

Routledge is an imprint of the Taylor & Francis Group, an informa business

British Library Cataloguing in Publication Data
Considering animals: contemporary studies in human–animal relations.
   1. Human–animal relationships. 2. Animal welfare. 3. Animal rights. 4. Animals in motion pictures. 5. Animals in art.
   I. Freeman, Carol. II. Leane, Elizabeth. III. Watt, Yvette.
   179.3-dc22

Library of Congress Cataloging-in-Publication Data
Considering animals / edited by Carol Freeman, Elizabeth Leane, and Yvette Watt.
   p. cm.
   Includes bibliographical references and index.
   ISBN 978-1-4094-0013-4 (hardback: alk. paper)
   1. Human–animal relationships. 2. Human–animal relationships—Philosophy. I. Freeman, Carol. II. Leane, Elizabeth. III. Watt, Yvette.
   QL85.C667 2011
   304.2'7—dc22

2011001594

ISBN 9781409400134 (hbk)

Contents

Part 3   Agency

List of Figures

Notes on Contributors

Philip Armstrong teaches in the English and cultural studies programmes at the University of Canterbury in New Zealand, where he is also the co-director of the New Zealand Centre for Human-Animal Studies (www.nzchas.canterbury.ac.nz). He is the author of *What Animals Mean in the Fiction of Modernity* (Routledge 2008) and the editor, with Laurence Simmons, of *Knowing Animals* (Brill, 2007). Philip is currently collaborating with Annie Potts and Deidre Brown on a book entitled *Kararehe: Animals in New Zealand Literature, Art and Everyday Life.*

Steve Baker is Emeritus Professor of Art History at the University of Central Lancashire, UK. He has been associated with the international development of the field of animal studies in the arts, humanities, and social sciences since the 1990s, and his particular interest is in the distinctive contribution made by contemporary artists to the development of this field of study. His publications include *The Postmodern Animal, Picturing the Beast*, and, with the Animal Studies Group, *Killing Animals*. His forthcoming book is *Art Before Ethics: Animal Life in Artists' Hands.*

Jonathan Balcombe studied biology in Canada before earning a PhD in ethology from the University of Tennessee. He has written many scientific papers and lay-articles on animal behaviour and animal protection. Balcombe is the author of *Second Nature: The Inner Lives of Animals* (Palgrave, 2010), *Pleasurable Kingdom: Animals and the Nature of Feeling Good* (Macmillan, 2006), and *The Use of Animals in Higher Education: Problems, Alternatives, and Recommendations* (Humane Society Press, 2000). Forthcoming is *Exultant Ark: A Pictorial Tour of Animal Pleasure* (University of California Press, March 2011). Balcombe works as an animal protection consultant in Washington, DC.

Marsha L. Baum, Dickason Professor of Law at the University of New Mexico School of Law in Albuquerque, New Mexico, teaches both law students and undergraduate students in the areas of property, ethics, animal law and animal advocacy. Baum's scholarship interests include animal law and weather and the law topics, which have intersected in discussion of animals in weather disasters. Her publications include *When Nature Strikes: Weather Disasters and the Law.* Baum is a member of the New York State Bar. She has degrees from the University of Rochester, Columbia University, and SUNY at Buffalo Law School.

Lucy Davis is a visual artist and writer on art, culture, and nature in Southeast Asia and Assistant Professor at the School of Art Design and Media, Nanyang Technological University, Singapore. She is Founding Editor of the publication series *FOCAS Forum On Contemporary Art & Society* and is published in *The*

DOCUMENTA #12 READER (Taschen); *BROADHSEET Art & Culture* (Australia); *Art Asia Pacific* (Sydney/New York); *Inter-Asia Cultural Studies* (Routledge) and *The Nordic Art Review* (Stockholm). Lucy is founder of the MIGRANT ECOLOGIES PROJECT, www.migrantecologies.org, concerning research into relations between culture and nature, Southeast Asian art and life. She is also Artist in Residence with Double Helix Tracking Technologies; a collaboration to trace "memories" of rainforest products via DNA timber tracking.

Carol Freeman is a Research Associate at the University of Tasmania and editor of the quaterly *Australian Animal Studies Group Bulletin*. Her work on visual representations of animals, bioethics, and the role of popular culture in wildlife conservation has appeared in publications as diverse as *Society and Animals*, *reCollections: Journal of the National Museum of Australia*, *Australian Zoologist*, and an essay collection, *Leonardo's Choice: Genetic Technologies and Animals*. Her book *Paper Tiger: A Visual History of the Thylacine* was published by Brill in 2010.

Elizabeth Leane is a senior lecturer in the School of English, Journalism, and European Languages at the University of Tasmania. She holds degrees in both literature and science, and is the author of *Reading Popular Physics: Disciplinary Skirmishes and Textual Strategies* (Ashgate 2007). Her research has been published in journals such as *The Review of English Studies*, *Signs*, *Ariel*, *Science Fiction Studies*, and *Polar Record*. In 2004 she travelled south on an Australian Antarctic Arts Fellowship as part of a project looking at literary representations of the Antarctic. Her current research focuses on narratives of human-animal encounter in Antarctica.

Tim Low is a biologist, consultant, and prize-winning author of six books. His most recent book, *The New Nature*, was praised by *Time* magazine and rated by *Who* magazine (Australia) as one of the books of the year. He writes for *Australian Geographic* and had a column in *Nature Australia* for 20 years. He divides his time between Brisbane and a cabin in a forest.

Jed Mayer is Assistant Professor of Victorian Literature at SUNY-New Paltz. His research focuses on the changing role of the nonhuman animal in the nineteenth century. He is currently at work on a book manuscript entitled *Scientific Dominion: Experimenting with the Victorian Animal*, which explores scientific, ecological, and ethical debates surrounding the nonhuman animal in nineteenth-century culture. Articles on cultural responses to evolution and vivisection have been published recently in *Victorian Studies, the Journal of Pre-Raphaelite Studies, Literature Compass*, and *Victorian Poetry*.

Kay Milton was, until recently, Professor of Social Anthropology at Queen's University Belfast, where she specialised in environmental anthropology. She conducted research for 20 years on environmentalism in the UK and Ireland, and briefly in Australia and New Zealand, focusing on environmental policy and politics, nature conservation, and human-animal relations. Her interest in environmentalism led to research on emotions as motivators of human action. She is the author of two books on environmental anthropology and numerous journal articles and book chapters. She now lives in New Zealand where she participates in practical conservation projects.

Stephanie Pfennigwerth has Honours degrees in Communications and Antarctic Studies, and an MA in English. She has worked as a researcher, writer, and editor for more than a decade, feeding her fascination for animals with trips to the Arctic, the Antarctic, and places in between. After a stint at the Australian Antarctic Division she swapped the ice for the outback, co-ordinating community education initiatives in the Shark Bay World Heritage Area, Western Australia. She won the 2009 Stearn Prize for natural history writing and is currently an assistant curator at the National Museum of Australia, Canberra.

Undine Sellbach is a writer, performer, artist, and philosopher based at the University of Tasmania. Her work explores the imagination, ethics, animality, and the unconscious. She is currently working on a project about scientific and creative attempts to imaginatively inhabit the lives of insects.

Helen Tiffin is the co-author of the influential text *The Empire Writes Back: Post-Colonial Literatures, Theory and Practice* (Routledge, 1989) and numerous other publications within postcolonial studies. She holds a BSc in science, specialising in zoology, and has increasingly applied the insights of her postcolonial research to human-animal studies. Her recent publications include (with Graham Huggan) *Postcolonial Ecocriticism: Literature, Animals, Environment* (Routledge, 2010) and the edited collection *Five Emus to the King of Siam: Environment and Empire* (Rodopi, 2007). She is currently an Adjunct Professor at the University of New England.

Yvette Watt is an Associate Lecturer in Fine Art at the Tasmanian School of Art, University of Tasmania. She is a practising artist and animal rights advocate, and issues surrounding the animal as a self-interested and sentient individual, as well as humans' ethical responsibility to animals, have informed her artwork for over two decades and lie at the core of her academic research. Her entry "Art, Animals and Ethics" appears in the recently published second edition of the *Encyclopaedia of Animal Rights and Animal Welfare* (Greenwood Press, 2009).

Wendy Woodward is a Professor in the English Department at the University of the Western Cape in greater Cape Town. Her research on literary representations of nonhuman animals culminated in *The Animal Gaze: Animal Subjectivities in Southern African Narratives* (Wits University Press 2008)—the first monograph in Human-Animal Studies in Southern Africa. Her next research project will focus on animals in the city—in fiction, the media, and photographs. She is also a published poet and her second volume of poetry *Love, Hades and Other Animals* (Protea 2008) engages with animals in Greek myth, colonial family histories, and daily experience.

Foreword

Jasper is an Asiatic black bear, also known as a moon bear because of the yellow crescent on his chest. He arrived at the Animals Asia Moon Bear Rescue Centre[1] outside of Chengdu, China, in 2000. Jasper was suffering from serious physical and psychological trauma, as are many of the bears who arrive there—some have missing paws, worn down teeth, or liver cancer. Jasper had spent 15 years living in a tiny, filthy "crush cage" on a bear farm, where his bile was milked with a rusty catheter and used in traditional medicine. But when I met Jasper, his gait was easy and smooth as he approached me. I fed him peaches and peanut butter and as he looked at me his eyes seemed to say, "All's well. The past is past, let's move on." Jasper has made friends with other bears at the Rescue Centre, plays in a hammock, swims in a water hole, exhibits trust and gentleness, and has an unerring ability to make other bears and humans feel at ease. He has a lot to teach us in terms of forgiveness, compassion, and empathy.

Bears, like humans and other animals, display different personalities. While attending a graduate program at Cornell Medical School, I staged protests against the school's use of animals for research and then returned to Washington University (in St Louis, Missouri) to study animal behaviour. During my years as an evolutionary biologist and behavioural ecologist at University of Colorado in Boulder, I was fortunate to meet and study domestic dogs, coyotes, gray wolves, Adelie penguins in Antarctica, and various birds. I came to the overwhelming (and obvious) conclusion that animals have not only emotional lives, but moral lives too. They exhibit wild justice. Animals have the ability to make judgments about the right and wrong ways to interact socially with other individuals. The evidence is very clear and it is being recognised more and more that animals exhibit emotional and moral intelligences. This much-needed paradigm shift in accepting that animals are sentient, emotional, and moral beings means that we must reassess our attitudes toward them and treatment of them. That is why Jane Goodall and I founded the organisation Ethologists for the Ethical Treatment of Animals.[2]

As an ethologist, I am aware that we must learn more about animals both as individuals and how they live in a group—what they feel, how their existence is bound to human desires and attitudes, and how what they experience affects their quality of life and survival. Animals *do* have a point of view on what happens to them. We need to know what will help Jasper and other animals who are used, abused, neglected, and marginalised, and do something about it. Indifference is deadly and mere words don't do much unless they inspire action.

---

[1]   See http://www.animalsasia.org/

[2]   See http://www.ethologicalethics.org/.

Attitudes towards animals are slowly changing and this book is an indication of that transformation. Many people are working "in the trenches." On one of my trips to China to visit Jasper and the other moon bears I worked with Jill Robinson, founder of Animals Asia, and a team of dedicated people at the Qiming Animal Rescue Centre, who tirelessly helped animals lost and injured in the earthquakes in Sichuan Province in May 2008. Scholars as well as those who work for animals outside of the ivory tower can have an impact transforming attitudes toward animals. Thus, the contributors to this book are involved in helping animals by drawing attention to various aspects of animals' lives in a scholarly context and by closely examining how they constantly and inevitably intersect with humans.

Thanks in part to my own research and that of people such as those we meet in this collection, new and exciting ideas about the importance of animals to humans are increasingly appearing in all areas of research. It takes courage to take a stand for animals in any discipline, to be an activist, as there is long-held resistance to anything but human priorities in mainstream academia. I am aware of the obstacles that have been overcome, the sacrifices that have been made, and the commitment required by the contributors and the editors to produce this book. In my travels I meet animal researchers from many disciplines and in various cultures, including many of those featured here. Some are well-established academics, whereas others are just beginning to get excited by the rewards and insights that come from focussing on animal-related topics. In either case, they write from the heart, as well as from the head.

One of the most encouraging aspects of this book is the presence of scientists writing alongside social scientists and humanists. For all the current rhetoric about the importance of interdisciplinary collaboration, such efforts remain relatively rare in practice. The scientists writing here—Jonathan Balcombe and Tim Low— are not writing scientifically, but rather using their research to reach out to a wider scholarly community. Elsewhere I have written about the importance of scientists acting as "concerned citizens," activists who make their work accessible to non-academics, and this is exactly what Jonathan and Tim are doing here. The emerging fields of conservation psychology and conservation social work[3] with which I'm increasingly involved at the University of Denver are good examples of how researchers and practitioners can collaborate to make the lives of animals better. Some of the contributors to this volume are working within these disciplines without even knowing it. Good for them!

Other contributors are also crossing party lines and reaching out to non-academic audiences. It is important for scientists to keep in mind that it is not only hard data that influences the ways in which people feel about animals and the natural environment. We also need to consider the impact of social and cultural factors, some of them quite concrete, such as the laws and statutes affecting animals that Marsha Baum discusses, and other intangible factors less easy to pin down,

---

[3]   See http://www.conservationpsychology.org/ and http://www.humananimalconnection. org/

such as the outpouring of public emotion in response to an individual dolphin described by Philip Armstrong. We need to ask, as Kay Milton does, how things such as children's books can help influence attitudes towards wildlife management and conservation, and, as several contributors note, the ways in which artists, writers, reporters, and filmmakers can use their venues to challenge our outdated assumptions about animals.

This book also highlights the importance of recognising the interdependence of the sciences and the humanities as well as the interconnectedness of the agendas of those individuals interested in animal protection and those more interested in environmental issues. As I have argued elsewhere, the staggering, sickening, and unprecedented loss of biodiversity we are currently experiencing (the anthropogenic "sixth extinction") is a form of animal abuse and also demeans us. Tim Low warns us against seeing humans as separate from the rest of nature, and points out that both human and nonhuman animals can affect the environment in ways that can be good and bad for species conservation. Jed Mayer identifies nineteenth-century connections between ecological concerns and animal rights that are relevant to current debates. Elizabeth Leane and Stephanie Pfennigwerth point to the inadequate ways in which recent films about penguins address the threat of climate change to these superb birds.

As I've indicated above, I have not always had a steady relationship with academia. I'm often frustrated by the reluctance of academics to move outside of their rigid disciplinary paradigms, by their tendency to talk amongst themselves with little regard for the broader impact of their work, by their emphasis (particularly in the sciences) on a supposed neutrality that is often a disguised politics. The contributors to *Considering Animals* recognise the frequent complicity of academic research with the inhumane treatment of animals. Jed Mayer's essay describes the way in which animal experimentation went hand-in-hand with the professionalisation of scientific research in the late 1800s. And Philip Armstrong similarly points out that academic research has for the most part looked with disdain on the kind of "sentimental" qualities—forgiveness, peace, trust, and hope—that I see as key to changing our relationship with animals and the natural world.

What impact, then, can a scholarly collection expect to have on the future of an abused animal such as Jasper the moon bear? It depends greatly on researchers' abilities to write and act as concerned citizens and as activists. It's not radical to care about animals and the environment. Indeed, it should be radical not to care and work for a better and more peaceful world! To this end, many of the contributors to *Considering Animals* examine concrete cases of animal suffering from around the world including canned hunting parks in South Africa, cat-culling in Singapore, and the abandonment of companion animals during Hurricane Katrina. They all write with the ultimate view of changing attitudes and actions toward animals among their colleagues and among the postgraduate students they supervise and the undergraduate students they teach. Ideas that begin in the seemingly rarefied atmosphere of the university before too long permeate society at large. Remember that campuses were once, and can again be, places of political ferment and activism. It was on a campus that my political campaigning for animal protection took public expression. In that case, I was protesting *against* the values of my lecturers and tutors. As Carol Freeman's essay on the dodo amply demonstrates, academic research and academic essays can equally challenge society's long-standing assumptions about nonhuman animals, and can lead students to do the same.

Finally, let's consider Henry and Jethro and the lessons we can learn from animal beings. At the Qiming Animal Centre, I met a dog I named Henry who had had most of his front leg lopped off by a butcher. Heather Bacon, the veterinarian at the Moon Bear Rescue Centre, took care of Henry and he came to trust people although he'd been so severely mistreated. Henry reminded me of my dog Jethro, whom I rescued from the Boulder (Colorado) Humane Society. Jethro came home with me, kept me healthy and happy and in many ways rescued me, and taught me many important lessons in life, including forgiveness and how it's possible to recover from untold traumas and to display empathy for others. People who live with companion animals know what a full life cats, dogs, and other animals can enjoy—that they have likes and dislikes, feel emotions, and deserve our respect, compassion, empathy, and love. Scientists and scholars in the humanities are coming to recognise these qualities and the importance of animals, as this unique and forward-looking volume convincingly shows.

Marc Bekoff

University of Colorado, Ecology and Evolutionary Biology
University of Denver, Graduate School of Social Work,
Institute of Human–Animal Connection.

Introduction

Carol Freeman and Elizabeth Leane

On an evening in 2007, in a small art gallery in inner-city Hobart, Tasmania, a crowd of about 20 visitors listened to a floor talk by Melbourne artist Deborah Williams. Her exhibition of etchings and engravings of dogs—silhouetted street dogs, dog shadows, lying dogs, lolling dogs, dark figures against grey or orange backgrounds—covered the walls of the gallery. The listeners stood in a ragged circle, framed by the relaxed, solitary, simply drawn subjects of the artworks, and heard talk of acquatint techniques, chemical hazards, and the artist's experiences. The large windows of the converted nineteenth-century post office provided a good view of the twilight sky and of the black dog who suddenly appeared during the talk, insistently and unmistakably barking to be let in. This was nobody's dog. That is, no human at the gallery that night was familiar with him. The curator let him in, the artist did not stop talking, and the dog entranced the listeners. He lay on the floor and listened too, head in the air, panting occasionally, and after a while he barked to the crowd. Did he want to leave? His body movements signalled he definitely did not—he was there to stay until the last visitor left and the gallery closed. It was an uncanny episode, an "animal moment."

Human lives and human histories are constantly punctuated by such moments—individual incidents and encounters that disturb our self-contained existences and bring into focus the complexities of our ongoing relationships with nonhuman animals. They are the raw material of the recent turn within both scholarship and public awareness that has been dubbed the Animal Moment. In 1998 geographers Jennifer Wolch and Jody Emel wrote of "witnessing the animal moment" in their edited collection *Animal Geographies: Place, Politics and Identity in the Nature-Culture Borderlands*. They asked: Why are animal-human relationships suddenly so topical and central to social theory? What political and intellectual purposes are served by studies of the "animal question"?[1] These questions have not been limited to the social sciences. In 2007 Marjorie Garber, author of *Dog Love*, an exploration of the relationship between dogs and humans, moderated a panel discussion on the place of animals in the history of science, natural history, evolutionary biology, literature, and the law, called "Animal Crossings." She too identified an "Animal Moment" that we are currently experiencing.[2]

Central to this Animal Moment has been the development of the scholarly field of human-animal studies. Originally undertaken at the margins of established

---

[1]   Wolch and Emel, "Witnessing the Animal Moment."

[2]   Garber quoted in Walker, "Scholars Probe Changing Legal, Cultural Status of Animals."

disciplines, human-animal studies has migrated into mainstream academia and dispersed into almost every area of the humanities as well as the social sciences, and formed inescapable links with the natural sciences. Interest in the animal has simultaneously emerged or developed in religious, public, and commercial spheres where there are theological debates and new laws concerned with the welfare of animals. Animal images and wildlife documentaries in new media, including online networks, film, video, television, and commercial advertising, have become increasingly popular. Video sharing websites such as YouTube often focus on human-animal relations and animals in the city, and there has been a preponderance of cartoons and books about animals in the late twentieth and early twenty-first centuries. With growing concern about the effects of climate change, pollution, and deforestation on animals as well as humans, practitioners and theorists in the sciences, social sciences, and the humanities are beginning to work together to find solutions to these global problems.

The incident of the black dog provides a useful example of this necessary enmeshing of disciplines. What brand of researcher could most profitably analyse and explain that particular animal moment? An art critic, historian, or theorist might best interrogate the complex relationship between the figures on the wall and the living subject who entered through the door. A zoologist could enlighten us about the dog's biological and anatomical nature, and its relationship to other breeds and species. An ethologist might provide informed speculation on the motivation behind the dog's own actions. A sociologist could shed some light on the reaction of the crowd standing in the gallery. A geographer would trace the dog's movements and trajectories and his place in the built environment. We might introduce a legal theorist, ethicist, or philosopher to argue the issue of the apparently ownerless dog's rights as they are currently legislated by humans. A cultural theorist and a psychologist could work together to unpick the powerful symbolism (for humans) of the "black dog." To give an adequate description of this animal moment—like all others—we would need a conference room full of academics from very different backgrounds, whose opinions would no doubt conflict at some points and converge at others, producing a rich, rewarding mix.

In the past, research has often focussed on animals as physical objects (in the case of the sciences) or as cultural artefacts, symbols, or actants (in the case of the social sciences and humanities). When a research project is brought under the rubric of "human-animal studies," however, its approach, emphasis, and aims change. Much debate has centred on the problematic first half of this term—"human-animal"[3]—but the second half is equally revealing. In her brief

---

[3]   Many researchers—such as those belonging to the British Animal Studies Network— use the term "animal studies."* We follow Ken Shapiro in adopting the term "human-animal studies" despite its obvious shortcomings (it is, he notes, "as incoherent as saying 'carrots and vegetables'"). Shapiro argues that it is "the best name at this time because it highlights the contradictions in current usage while retaining the emphasis on human-animal relationships"—which "animal studies" does not. *Human-Animal Studies*, 5–6.

discussion of the rise of "studies" in academia, Garber notes the term, as it first emerged several decades ago, applied to fields that were "precisely *not* disciplines; they were interdisciplines, nexes of overlapping interest." Originally indicating a geographical or historical focus (American studies, medieval studies), the term later took on more explicitly political resonances (gender studies, postcolonial studies), making it for some "a suspect piece of jargon."[4] By adopting the term "human-animal studies," the field places itself unashamedly in this tradition, endorsing a steadfastly anti-disciplinary and explicitly political outlook.

This anti-disciplinarity means that human-animal studies is defined more by the subject of its analysis—what Philip Armstrong and Laurence Simmons identify as "the cultural, philosophical, economic and social means by which humans and animals interact"—than any shared methodology.[5] In a recent policy paper defining the field, Ken Shapiro argues that human-animal studies interrogates "the various ways in which nonhuman animals figure in our lives and we in theirs." It investigates the "impressively variable forms of bonds, attachments, interactions, and communications" between human and nonhuman animals—relationships that "can be symbolic, factual or fictional, historical or contemporary, and beneficial or detrimental to one or both parties." A feature of the field is that a human-animal studies scholar *reflects on*, as well as describes, the "limitations and complexities" of these relationships.[6]

The field has already generated a significant body of scholarly literature that impacts on current approaches, generating a loose consensus of political views and a shared understanding of relevant concerns and ideas. Collections of essays have had an influential role in the development of the field, with their format enabling the multidisciplinarity that human-animal studies requires. While some titles, such as *Animal Geographies*, usefully show how a particular (if flexible) disciplinary approach can contribute to the field, and others, such as *Victorian Animal Dreams*, direct themselves at a defined readership by taking a historical focus, the majority are explicitly inclusive when it come to disciplinary framework. Armstrong and Simmons emphasize, for instance, that their collection *Knowing Animals* "is a kind of mixed human-animal habitat into which diverse species have been introduced, to intermingle and interbreed, appropriately or not."[7]

However, the potential for multidisciplinarity that the essay-collection format allows has yet to be exploited to its full extent. The disciplinary habitats of most recent collections do not, in practice, extend far past the humanities: the occasional social scientist peers out from behind the foliage, but very rarely a biological or physical scientist. There are evident reasons for this, in addition to the practical and institutional obstacles to crossdisciplinary conversation and publication. The

---

4   Garber, *Academic Instincts*, 77.

5   Armstrong and Simmons, *Knowing Animals*, 1.

6   Shapiro, *Human-Animal Studies*, 1–13.

7   Wolch and Emel, *Animal Geographies*; Morse and Danahy, *Victorian Animal Dreams*; Armstrong and Simmons, *Knowing Animals*, 3.

humanities are very belated in including animals within their purview; they are, as Cary Wolfe writes, "now struggling to catch up with a radical revaluation of the status of nonhuman animals that has taken place in society at large."[8] But human-animal studies will be impoverished if it sees the humanities or the social sciences as mere complements to the natural sciences, sitting alongside them and adding to or challenging their separate knowledges.

One of the objects of *Considering Animals* is to show the benefit of widening the disciplinary mix. In addition to researchers working within literary and visual arts, philosophy, social anthropology, media studies, cultural studies, and law, two well-established scientists who think outside the conventional scientific square—Jonathan Balcombe and Tim Low—have contributed to the volume and we asked a third, Marc Bekoff, to contextualize the collection in a foreword. Two of the literary scholars, Helen Tiffin and Elizabeth Leane, also have scientific qualifications. Multidisciplinarity of this kind is a step towards a deeper interdisciplinarity, in which scientists, social scientists, and scholars in the humanities collaborate intellectually not only in the publication but also the research process. What is required is an integrated approach, in which scientists work together with their colleagues in the humanities to achieve more liberal and sensitive interactions with and views of animals, taking into account researchers' own relationship with their subjects. This would mean rethinking the conventions of their respective disciplines and re-examining and reforming their attitudes toward the behaviour, appearance, and habits of animals and their relationships with us. This integrated approach is one that the field is only now beginning to feel its way towards, and it is through volumes such as *Considering Animals*, where essays from different disciplines are arranged under themes of primary importance to animals themselves, that researchers from seemingly disparate disciplines can begin to identify links and resonances within their work.

While *Considering Animals* aims to widen the scope of human-animal studies collections, its title—which echoes in its grammatical structure recent collections including *Representing Animals*, *Knowing Animals*, *Killing Animals*, and *Figuring Animals*—signals its continuity with existing scholarship. Like those collections, we use the present participle to indicate progressive movement—as an acknowledgement that all understanding of human-animal encounters is partial, contingent, and in-process. *Considering Animals* also puns deliberately on the participle, inviting contributors and readers to *consider animals* in several related ways. To begin with, we want to interrogate what we humans do when we are in the act of considering animals. Thus the first group of essays asks: How do we represent nonhuman animals to ourselves? Simultaneously, we need to keep constantly in mind our practical and ethical relationship with animals: how considerate are we of species beyond our own? The second group is primary focussed around this question. And, most challengingly, we acknowledge how animals themselves have been increasingly shown to be capable of considering each other as well as us—

---

8   Wolfe, "Introduction," xi.

that they think, communicate, and exercise agency. This is something particularly relevant to the final group of essays; but because each interpretation of the term "considering" informs the others, there is often an overlap between the concerns of the essays in each section.

These considerations bring us back to both the specific animal moment with which we began, and the broader intellectual Animal Moment we are currently experiencing. The image evoked earlier of a room full of researchers—visual artists, zoologists, behaviouralists, sociologists, geographers, lawyers, cultural theorists, psychologists—puzzling over the incident of the black dog risks leaving out of the equation the one subject integral to it. What role do the dog's own actions have in the incident and how can we engage with this question without forcing our own human perspectives—with whatever good-willed intent—upon the dog? Similar questions have plagued other evolving multidisciplinary studies. Modern feminist studies, for instance, after its energetic beginnings, started to ask which subjects—which women—it was eliding. Who was speaking for whom and whose differences were being erased by this act? These questions are particularly pressing in human-animal studies, which must continually ask how we deal with the non-speaking subject, or even what it means to privilege the category of human speech. As Wolfe suggests in his introduction to the collection *Zoontologies*, the relationship between language and subjectivity is at the heart of current debates within the field.[9] Some of the contributors to this volume—such as Helen Tiffin, in her essay "The Speech of Dumb Beasts"—deal with this issue head-on; all were asked to keep, in whatever way they chose, the material, acting animal at the centre of their analysis.

Our invitation to a multidisciplinary group of researchers—some leaders in their field, others emerging scholars—to further the process of "considering animals" provoked a wide range of responses. Some deal with recent or current events in which, as is often the case, animals have been forgotten, little understood, or largely invisible to scholars. Others tackle animal subjects that are ignored by academia because they are considered trivial, or still fall foul of formulaic discourses or simplistic or prejudiced theoretical binds. The authors of these chapters rethink representational strategies, re-examine media portrayals, and question the meaning of words that have been applied to animal representations, such as "sentimental." Long-held and traditional opinions and research about animals are interrogated. In all of these cases it is the animal, not the human, who is kept at the centre of the text. In this sense, this volume challenges the oft-repeated notion that animals are "good to think" with—that they tell us mainly about ourselves—for its contents are focussed firmly on the nonhuman.

The essays in Part 1, "Images," consider textual animals whose presence in a film, artwork, or other representational form has previously been marginalised or underestimated, or given only symbolic significance. Many images have the capacity to challenge assumptions and preconceptions. However, only a very few

---

[9]   Ibid., xviii.

contemporary artists working with animal imagery see their work as being directly shaped by a commitment to the cause of animal rights. Steve Baker looks at recent work by three of them: US-based Sue Coe, UK-based Britta Jaschinski, and New Zealand-based Angela Singer. Drawing on his recent interviews with each of these artists, he considers how their very different practices present productive (but often unexpected) strategies for undermining conventional approaches to animals. For each of the artists, their varied media, materials, and presentational strategies are the means by which they add something that takes the viewer beyond "photographs of ... the killing floor." That something, Baker argues, is—to borrow the language of biosemiotics—"a difference that makes a difference."

Animals are at the centre and yet simultaneously missing from two bestselling films discussed in Elizabeth Leane and Stephanie Pfennigwerth's essay. Emperor penguins are the ostensible stars of the most successful nature documentary ever made, *March of the Penguins*, but the film's unexpected embrace by conservative religious groups meant that it was not the birds themselves, but the degree to which their actions paralleled human behaviour (particularly parenting methods), that drew attention. The film's mythologically inflected construction of a penguin colony, which emphasizes its timelessness and eschews mention of either evolutionary adaptation or human interference, means that the threat of global warming to the birds' well-being is entirely ignored. The highly successful animated film *Happy Feet*, released the following year, brought home an environmental message, which seemed a corrective to *March of the Penguins*. Yet Leane and Pfeninngwerth show that the film's narrative conservatism prevents a coherent moral from emerging. The two films are, they conclude, "united in their inability to go beyond the symbolic animal."

Another animal documentary—Werner Herzog's *Grizzly Man*—is the focus of the next chapter, by philosopher and writer Undine Sellbach. Herzog's film deals with the relationship between amateur filmmaker Timothy Treadwell and a group of wild grizzly bears whose Alaskan wilderness home he regularly shared. Herzog incorporates selected sections of Treadwell's own footage, refusing, however, to include an audio recording of the fatal attack on Treadwell and his girlfriend by a grizzly. Sellbach treats the traumatic Animal Moment around which the film is structured as indicative of the distress which interaction between the self and other can entail. She thereby explores a question that lies at the heart of human-animal studies: what really occurs in encounters between human and nonhuman animals, and how can this be conveyed?

Wendy Woodward turns our attention initially to another violent animal moment: the "ceremonial" sacrificing of a bull in another medium—South African print journalism. Newspaper reports of the death of the bull and other encounters between humans and domestic or farm animals tend to individualize the animal in question. The impact of this is not, Woodward argues, to alter human assumptions about animals generally but, rather, to render the newsworthy animal exceptional and thus to remove it from the routinely accepted treatment of nameless animals reared for consumption. This strategic slippage does not apply, however, to media depictions of wild animals, which encourage the reader to treat the individualized subject as representative rather than exceptional. This creates a double standard in which the treatment of wild animals evokes strong emotions while the routine killing of farm animals goes unquestioned.

To what extent does the appearance of an animal determine human attitudes towards it, and can this process be deliberately manipulated? Kay Milton's chapter focuses on a group of animals who have persistently resisted human attempts at their eradication. An introduced species in New Zealand, possums pose a significant threat to its native wildlife, and their destruction is thus advocated by environmentalists and implemented in public campaigns. Yet this is an animal that fits supposedly innate human assumptions about "cuteness." Milton (an anthropologist) explores public attempts to counter this response—to make the possum "uncute"—in order to render its wide-scale destruction more acceptable to New Zealanders.

Part 2, "Ethics," explores our practical and ethical relationships with animals, taking into account their capacity for pain and pleasure. Ethologist Jonathan Balcombe discusses recent studies of the everyday experiences and feelings of nonhuman animals. Mice empathize with familiar mice who are suffering, capuchin monkeys have a sense of justice, and rats accustomed to being tickled will come running for more. Unfortunately, as knowledge of animal sentience advances, our treatment of them lags further behind: we kill more than 50 billion land animals yearly. Balcombe asks us to put aside our preoccupation with intelligence and to recognize that animals' pain and pleasure are akin to ours—and their will to live is just as strong.

Jed Mayer's essay introduces the pressing subject of ecology. In the twenty-first century, public debates on animal rights and ecological issues rarely intersect, yet when these discourses were first beginning to emerge in Great Britain during the later years of the nineteenth century, there flourished a more unified conception of the nonhuman world which emphasized the link between the rights of animals and the value and vulnerability of the earth's dwindling green spaces. Mayer investigates the rhetoric of vivisection, where the experimental animal served as an embodiment of the natural world which could be penetrated and controlled by the heroic man of science, and considers the ways in which the laboratory animal represented the larger nonhuman world in the late nineteenth century. He suggests continuities between ecological and animal rights concerns that might productively inform current debates and events.

The devastation of the city of New Orleans by Hurricane Katrina—a widely reported event—and its impact on individual relationships between human and animals is the subject of Marsha Baum's essay. Baum examines legislation that deals with the treatment of companion and service animals during natural disasters. The tremendous anguish generated when humans were forced to abandon their animal companions during the Katrina crisis led to legislative changes which ostensibly took animals "into account" to a greater extent than previous legal instruments. Baum argues, however, that it is primarily human, rather than animal, needs that are considered in these new laws. By treating the Katrina disaster as an animal moment, Baum asks us to rethink our view of animals' intrinsic value and their self-interest in emergency situations.

The section ends on a personal note, with artist Yvette Watt writing about her own art practice and her role as an animal rights activist in relation to the work of a growing number of other artists who have taken animals and animal-human relationships as their subject matter. Watt sees cause for concern in the manner in which animals are presented by many artists and curators. She argues that despite a general postmodern avoidance of the animal as symbol in these recent shows, animals are nonetheless too often present as generic signifiers for the natural world, rather than being presented as individual, sentient, and self-interested beings. This, she suggests, results in the animals becoming marginalized, allowing the artists to avoid addressing the broader ethical issues surrounding the way humans interact with animals. Watt's concerns reflect those of many of the contributors to the second part of the volume, who document the ongoing vanishing act in which animals—or particular kinds of animals—are effectively erased in favour of human interests.

In Part 3, "Agency," our attention is turned from human consideration for animals, to animals as considering beings themselves. Helen Tiffin's essay draws attention to human attitudes towards animals' capacity for speech. The recognition of speech, she points out, has been crucial in "marking" and dividing humans from animals, as well as creating racial hierarchies within the human communities. One rare space in which the concept of a speaking animal is allowed is the novel; yet, outside of children's literature, fictional animals have usually served comic or satirical purposes, in which animals are "merely masks" for humans. Tiffin maintains, nonetheless, that imaginative literature is capable of giving voice to animals in ways unavailable to other discourses. To this end she examines a number of contemporary writers who undermine generic conventions in order to represent speaking animals "as *animals*" rather than substitute humans.

Carol Freeman focuses on the human erasure of animal agency in a well-known extinction narrative. Visual images of plump, waddling dodos are everywhere in Mauritius, the country they once inhabited. Yet the exact appearance of this extinct bird is by no means certain, nor are the circumstances of the species' disappearance. Recent evidence based on a late twentieth-century scientific reconstruction of a dodo skeleton, however, suggests that dodos were "lithe and active." Freeman re-examines early visual and verbal impressions of the dodo, challenges David Gooding, who refers to this animal as one that "allowed" itself to be slaughtered, and explores definitions of agency. She not only reflects on transformations in human-animal relations over time, but also demonstrates what placing an animal at the centre of a discussion actually involves.

A very different animal narrative—that of Opo, a wild bottlenose dolphin who made almost daily visits to a popular swimming beach in New Zealand—provides an exemplary focus for Philip Armstrong's examination of the structures of feeling pertaining to cetaceans in the second half of the twentieth century. The many retellings of Opo's story, Armstrong notes, share a focus on the dolphin's "agency and personality." However, as with the speaking animals of literature that Tiffin discusses, the granting of subjecthood to a wild animal is often considered "sentimental." Armstrong situates this story within the context of other significant human-cetacean encounters, including that of Paul Spong and the Orca Skana during the 1960s; and those of Greenpeace protesters and Russian whaling ships in 1973. His analysis of such narratives demonstrates that human-animal relations are as heavily influenced by shared emotional states as they are by ideology and economics, but all too often the "structures of feeling" that operate between humans and other species are ignored or dismissed. Changes and continuities in the cultural mood that governs how animals can be understood provoke Armstrong to a reconsideration of what we mean by the term "sentimental."

Lucy Davis alerts us to the government-sponsored treatment of animals in a disaster, the outbreak of Severe Acute Respiratory Syndrome in South-East Asia in the first part of 2003. She demonstrates that the street cat, with its loud, unpredictable, and uncontrollable habits, has been strikingly resistant to government ideology as well as to the practical efforts to make it disappear. Davis has elsewhere written about the way in which Singaporean street cats became demonized during the crisis, with the Singapore People's Action Party government projecting onto the cat its obsession with hygiene and surveillance, as well as assumptions about gender, race, and class. Here, however, she is more concerned with experiences of the material animal than its symbolic treatment by humans.

In the last chapter, biologist and environmentalist Tim Low looks at the big picture, particularly the concept that humans play a central role in the earth's ecosystems. He argues that this kind of thinking overvalues human input and ignores nonhuman involvement, which often transforms the environmental changes that humans make. Low points out that these changes can be for better or worse as far as habitat conservation is concerned and better or worse for us: birds, for example, spread far more weeds than humans do. While the notion of "harmful humans and innocent nature" may be useful rhetorically for "advancing nonhuman interest," he argues, "it ultimately limits our thinking." We need to avoid simplistic binaries and accept nature as both an active agent and a multiplicity of which humanity is merely one part.

***

The scope of the topics covered in this volume gives some indication of the possibilities for further research and scholarship in human-animal studies. We still know and understand relatively little about nonhuman animals outside their native habitats and their roles as domestic "pets." To return to the gallery once more: even such simple questions as why the black dog came to the exhibition, where he lived, what he felt and thought about his experience there, what he was trying to say, and how he knew when to leave, reveal that the effort to understand animals will take more than only scientific or only humanistic enquiry. And as

Fig. I.1   Deborah Williams. *I imagine your subjectivity as both rich and clearly different*. Roulette, angle grinder and intaglio drypoint.

far as human-animal relations go, there are other questions. What role did the humans at the gallery play in this particular dog's life? Why did they react as they did? How were they influenced by their shared concern about animals? For a nonhuman animal presence elicited a profound disintegration of conventional behaviour at this gathering, signalling that the future of human-animal studies is more than the theoretical, academic, and legal investigation *of* animals; it requires us to more fully include them as presences in our analyses. It is time to let the dogs out—or rather, *into* the discussion.

PART 1
Image

Chapter 1
Contemporary Art and Animal Rights

Steve Baker

In an essay first published in *Parallax* in 2006 that includes discussion of artist Sue Coe's book *Dead Meat*, Cary Wolfe asks that given what Coe sees as the ethical function of her own art (a furthering of the cause of animal rights), "why not just show people photographs of stockyards, slaughterhouses, and the killing floor to achieve this end? To put it another way, what does art *add*?"[1] What does art add? Wolfe does not go on to answer this question directly, but it is a highly pertinent question that sets the broad agenda for the present essay.

Only a very few contemporary artists working with animal imagery see their work as being directly shaped by their own commitment to the cause of animal rights. This essay will briefly consider some of the recent work and concerns of three such artists: US-based Sue Coe, who is primarily associated with graphic work for the printed page; UK-based Britta Jaschinski, whose medium is photography; and New Zealand-based Angela Singer, who recycles taxidermy and other animal materials. My intention is not at all to offer a nascent *theory* of animal rights art, but rather to present a preliminary comparison of how and why these particular artists see their diverse presentational strategies as a necessary means of taking the viewer beyond Wolfe's "photographs of ... the killing floor." All quotations from the artists, unless otherwise indicated, are drawn from my interviews, conversations, and correspondence with them.

***

In *Chaosmosis*, Félix Guattari makes the striking claim that "The work of art, for those who use it, is an activity of unframing, of rupturing sense."[2] While this should not be regarded as setting out in advance my own view of the work and ambitions of Coe, Jaschinski, and Singer, it does offer a point of reference alongside which their work might be considered.

I will take Sue Coe first, and focus on her 2005 book *Sheep of Fools*. Produced collaboratively with the writer Judith Brody, and named as "Nonfiction book of the year" in the 2005 People for the Ethical Treatment of Animals (PETA) Progress Awards, it traces aspects of the historical development of the present-

---

[1]   Wolfe, "From *Dead Meat*," 99.
[2]   Guattari, *Chaosmosis*, 131.

day phenomenon of the live export of sheep. It is Coe's third book-format work to deal directly with animal rights issues. Like the earlier books *Dead Meat* and *Pit's Letter*, it incorporates drawings, paintings, and works in other media by Coe that were seldom made specifically as illustrations for those publications, and weaves them into a narrative in conjunction with the accompanying text. But while *Dead Meat* was in effect an extended diary based on Coe's travels around North American factory farms and slaughterhouses, and *Pit's Letter* was the story of one fictional dog's experience of (and reflections on) a variety of forms of incarceration, experimentation, and abuse, *Sheep of Fools* presents an ambitious and expansive (albeit highly unconventional) historical narrative that is packed into little more than 30 pages. Subtitled *A Song Cycle for Five Voices*, it takes the reader through the "Song of the Medieval Shepherd," "Song of the Venture Capitalist," "Song of the Modern Shepherd" (Figure 1.1), "Song of the Trucks and the Ships" (the longest section by far), and finally "Song of the Butcher."

The book's double-page spreads have been ably described by Jane Kallir as "masterful compositions that combine compassionate observation with dramatic narrative sweep."[3] In many cases they incorporate Brody's rhyming couplets: "A wooly tale of avarice and one that's seldom told: / how merchants spun their fortunes as they turned our fleece to gold." Coe is absolutely clear that the book's audience "will be activists." She says: "Really this work is directed primarily to activists ... it gives them the impetus to increase their desire, *the* desire for social change." She is equally clear that this is something that art can do, and that *her* art can do. In an earlier interview for the *Los Angeles Times*, one of her "five tips" for artists at the start of their careers had been: "Before art can be a tool for change, it has to be art."[4] I suggested to her that this reflected a strong conviction about contemporary art's potential influence and importance, and she acknowledged: "It's important to me, and it's a voice that worked."

Published by Fantagraphics Books—a name associated principally with alternative comics and graphic novels—*Sheep of Fools* is tagged "A BLAB Book," and it shares the same large square page format as the *BLAB!* Comics to which Coe contributed several eight-page illustrated storylines in the mid-2000s. In her interview for the illustration magazine *3 x 3* shortly before the book's publication, she noted, "I tend to work sequentially, in a mode that I think of as reportage, or visual journalism," and said of other art she admired: "The most exciting work for me today, is sequential art, stories directly observed in cartoon and book form."[5] Publishing in *BLAB!*, where no restrictions were placed on the form or content of her work, she found a freedom that had generally been unavailable to her in her work for more mainstream publications, and one of the first things she emphasised in our conversation about *Sheep of Fools* was, "that's the absolute bottom line of my creative processes—I have to have freedom."

---

3   Kallir, "Sue Coe: *Sheep of Fools*," n.p.
4   Vaughn, "Staying True to Art."
5   Anon., "Sue Coe Interview."

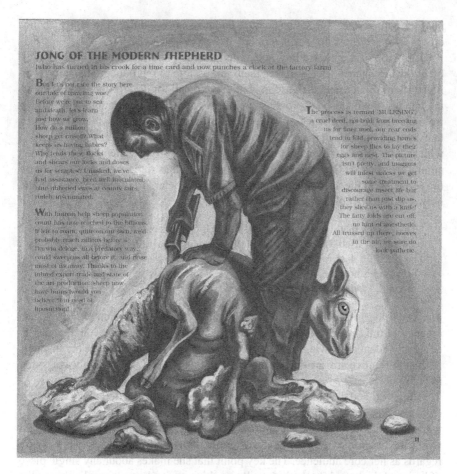

Fig. 1.1   Sue Coe, *Sheep of Fools*, 2005, p. 11. Courtesy of the artist and Galerie St. Etienne, New York.

The book's main focus is on the nature and the sheer scale of contemporary live export operations. Coe says of the individual drawings that preceded the book, "What started me on the series was I saw one little newspaper cutting"—about a ship that exploded—"where it said 'one life was lost,' but they didn't count the sixty-five thousand sheep as lives." She soon enough discovered that the ships' crews were non-unionised labour, uninsured, and untrained: as she says, "Twenty two men with eighty thousand animals, how does *that* work?"

She spoke of "fascinating bits and pieces that I wanted to put together to make vague sense," and acknowledged that it could easily have become "a bit preachy" if done entirely through language. But the written word is of course not the only alternative to her distinctive drawing style. Thinking very much of animal rights videos of slaughterhouses and the like, she describes her own chosen medium in the following terms: "the subject matter I address is very difficult to look at in any other form ... almost impossible to look at."

The scale of the book's highly detailed double-page spreads—10 inches high by 20 twenty inches wide—makes them impossible to reproduce satisfactorily as illustrations in an essay such as this, but they do call for description. And some of the images of which she is proudest are largely text-free. One double-page spread from the "Song of the Trucks and the Ships" section, for example, shows two ships that are about to pass in the night. One, the *Al-Yasrah*, has deck upon deck stacked with sheep, their faces looking out into the night, as two crew members hurl a dead animal overboard. The other, the *Princess Cruise*, has three couples waltzing on the foredeck to the accompaniment of a lone trumpeter. The ships' names are the only text in the whole spread. Coe remarked:

> That's one great drawing. That's the full Sue-Coe-ness ... And it's not a great drawing in itself, but it's got exactly the content I want ... because nothing's invented ... Just to put those two things together—that makes me happy—because it's not a lie, it's not an artifice, yet it's obviously stylized but it's based in truth ... One's a luxury cruise and one's a journey to death.

The key thing is directness, or what she calls "keeping the vision *mobile*." "The urgency of the content has to overcome any art-making process, even though I'm obviously adept at drawing," she says. "I'm always trying to sabotage my instincts ... Just to remove any artifice is an art in itself, and that's what I'm always trying to do."

The effort and the uncertainty is evident in a telling distinction that Coe draws, regarding her work as a whole, between *work that sells* and *work that works*—work that sells through the New York gallery that represents her, and work that works in the sense of reinforcing the determination of the animal activists who she regards as her core audience. The key point that she makes about any single piece of her work is that while she is actually working on it, she cannot reliably predict which of those two categories it will fall into: "You can't, you can't," she insists.

Juxtapositions of human and animal circumstances are found throughout *Sheep of Fools*. The page that follows the ships passing in the night shows the fate of the *Farid Fares*, a decrepit ship on which 40,000 sheep burned, as the hapless crew rowed away in lifeboats (Figure 1.2). I asked Coe if there was an intentional echo here, in the depiction of the falling sheep, of those indelible images of humans jumping from the Twin Towers on 9/11: "Yes, definitely," she said. "But the ship is the high-rise building, that's what it is, and they would have been jumping out. It wasn't like, I invented it; that would have been true." Other resonances crowd in, too, with Brody's text at the upper right of the page including phrases like "10,000 sheep packed Auschwitz-tight."

Coe describes her collaboration with Brody in very positive terms: "Once we started researching, it's endless and it's fascinating and that's the joy of it." Brody came up with the idea of the "song cycle," and her "doggerel—she calls it doggerel" is something that Coe very much likes. This *pleasure* in the actual

Fig. 1.2   Sue Coe, *Sheep of Fools*, 2005, p. 24. Courtesy of the artist and Galerie St. Etienne, New York.

making of an artwork that addresses such distressing events may surprise some readers, but it is an important aspect of Coe's ability to understand and effectively to handle the skills of her craft, as was her comment, quoted earlier, about "always trying to sabotage" her instincts, her *facility*. It also goes along with recognition of the limits of her skills. She has never felt entirely satisfied with the balance of image and text in the design of her books: "It's a failure in every single book I've done," she suggests. Of one particular artist she admires precisely for the ability to use "the text and the image as one," she says: "But he's a Buddhist from Japan; I'm a white working-class woman from Hersham … I can't do that same flow."

Coe admires flow and skill wherever she finds them. Of the sheep-shearer (Figure 1.1), for example, where—as in many of her images—the folded forms of shearer and sheep echo and cross each other in a deliberate and sympathetic manner, she says:

> This one, the sheep shearer, I drew hundreds of pictures of him. And sheep-shearing is such an art ... Their backs go very quickly. The hard part is not the shearing, because they do that flawlessly, with no nicks and cuts, in thirty-two strokes. But the flipping of the sheep is very hard on the back, and they do thousands of those sheep a week, shear them, and it's completely painless when they're good ... This is an art form that hasn't changed in thousands of years, and they're great at it.

It is an image that reflects well enough her working ethos: "quiet observation without projection." But even when other pages show more gruesome imagery—such as, on the page to the immediate left of the sheep-shearer, her fierce depiction of the horrific practice of "mulesing" the sheep (using shears to cut the fatty folds of flesh from around the trussed sheep's buttocks to prevent the disease flystrike)—Coe's aim is never simply to condemn those caught up in an industry that creates and permits those practices: "There's a neutrality in how I depict the human beings, I hope."

Aiming to be true to what she sees, and to address "social justice issues," will not always make her work popular, even among the activists she sees as her key audience. But as she says, "they don't have to *like* the work, in fact, many activists would prefer the cute puppy with the halo, you know, because they're regular people, they don't always want the horrible truth of what happens to animals."

***

Britta Jaschinski's award-winning photographs have been widely reproduced—on the covers, for example, of books such as Randy Malamud's *Reading Zoos* and Kalof and Fitzgerald's *The Animals Reader*, to name only two. The photographic series on which she has been working for the past few years carries the evocative title Dark (Figure 1.3). Although, like Coe, Jaschinski makes and exhibits individual images, she acknowledges that her books are probably the means by which her work reaches its widest audience. Following *Zoo* in 1996, and *Wild Things* in 2003, Dark is her third and (at the time of writing) as yet unpublished book project, which I saw in one of its several dummy versions. The photographs are all landscape-format, and most of them are indeed dark. That impression was reinforced in the dummy I saw, where in quite a number of the double-page spreads one of the two pages was a blank, black page—a strategy that had also been used in *Zoo*.

Unlike Coe's work, which invariably shows human and nonhuman animals locked together in circumstances that are seldom comfortable, the human is generally absent from or only indirectly present in Jaschinski's work. The photographs in *Zoo* admittedly often include details of the human-made architectural incarceration of the depicted animals, and the almost incidental presences of zoo visitors also appear in a couple of shots. In *Wild Things*, evidence of human degradation of the environment in which other creatures must also live—a central concern of the

Fig. 1.3   Britta Jaschinski, *Tiger*, 2006, photograph from the Dark series.
                Courtesy of the artist.

book—is apparent in many of the photographs. The other human "presence" is the
text that runs as a single broken line through that book, irregularly interspersed
with its many photographs, which takes the form of a letter from the artist herself
(printed in the style of lowercase typescript) that begins "dear animals," and ends
"your devoted friend." Along the way, it includes the sentence "human existence is
purely accidental," which gives a sense of her perspective on the issues.

    Jaschinski describes the Dark series as "much more abstract" than the
photographs in her earlier books, and hopes that there is some degree of progress
"in terms of the way I see animals and what I manage to communicate about
them." Hers is the difficult process of, as she puts it, "waiting for something to
photograph which nobody else has seen before." Even when she has managed to
do so, it may not immediately be clear to her. There's an ongoing process of editing
and re-editing. One absolutely extraordinary photograph of an orangutan is a case
in point. A lower jaw, an upper lip, a forehead, and part of a left shoulder are just
discernible as isolated clusters of fine diagonal lines of light sweeping from left to
right across the otherwise black image. The impression is of the animal's forward
movement, but it could equally be a retreat, or just as easily the camera's own
movement. The orangutan was not part of the initial sequence she had envisaged
for the dummy, but "suddenly I feel like I want this to be part of it," she says.

    The rhinoceros (Figure 1.4), too, now "needs to be in there." It "wasn't in
at the beginning, in fact it came in, and went out, and then it came back in."
Why? "It looks so incredibly lost," she says—*lost* being a word she constantly

Fig. 1.4   Britta Jaschinski, *Rhinoceros*, 2006, photograph from the Dark
              series. Courtesy of the artist.

uses to describe the animals photographed for this project. I asked how, from her
point of view, this particular animal's image differed from that of the rhino in her
earlier book *Wild Things*. There, in sharp Dürer-like focus, the sepia image of the
animal in profile is seen against what looks like (but is not) a blank photographic
studio backdrop. Appearing towards the end of the unpaginated book, it is one of
a sequence of pictorial interruptions over several pages of a line from the artist's
"letter" that has her saying to the depicted animals: "I hope you don't mind me
telling you that your strength and craving for an unrestricted existence make
you quite anachronistic." Those last four words appear on a right-hand page; the
rhinoceros photograph is on the left. Jaschinski's answer to how the two rhino
photographs differed was to say that the one in *Wild Things* "conveys a huge
amount of power"; the one in Dark (Figure 1.4) is the opposite. Here, she says,
"I'm trying to see it from the angle of the animal rather than the viewer ... and
that's so hard." "It looks so awkward," she continues; "what a weird picture ...
We never *really* look at animals, we look at animals the way we choose to look at
animals."

It is a matter of figuring out, she suggests, "what needs to be said next"—not in
terms of her own aesthetic ambitions, but rather in terms of the relation of nonhuman
animals to the state of society and the state of the planet. But that, of course, will
necessarily involve an artist such as Jaschinski in aesthetic decision-making, and
she is quite open about being dissatisfied with aspects of her earlier work. Of *Wild
Things*, for example, she says with a laugh: "It's such a mess, that book, isn't

it?" She also distinguishes her approach from that of certain other photographic traditions: "The more information an image has, the less interesting it is for me … That's why I'm not interested in a lot of wildlife photography." Audiences, she suggests, tend to trust what they've seen before, what's comfortable—whereas, she says, "I have absolutely no intention of glamorizing wildlife." She expands on the point: "I think there's nothing worse for a viewer than walking round in a gallery or looking at a book where everything's like, *easy*, you turn the page, you walk around, and it's *easy*. I always want to create a kind of challenge for the viewer … I think I haven't quite done that enough yet."

In saying of her work "I don't want to take a picture of anything people know," of course, she sets the bar *very* high for herself. But as much as anything, it is her conviction that "It *is* about what we don't know, about the animal" that justifies her hope *not* to be seen as "somebody who preaches." "I think it needs to be said," she adds, "that we will never know."

***

Angela Singer's uncompromising taxidermic constructions (Figure 1.5) have won her friends in some unexpected places. In an entry on "Lady Lavona's Cabinet of Curiosities" online blog, for example, she is described as "Artist, animal activist, goddess!"[6] I suspect that Singer herself would not particularly care for the accolade, but the growing recognition of her work within and beyond New Zealand is welcome and long overdue.

Like Jaschinski, Singer's conviction is that "people need to see animals in a new way," and that "artists can provide the new visual language." The artist's role is to "shock the viewer into a new way of seeing and thinking about the animal." Of an early work, *Portrait of a Naturalized New Zealander*, from 1999–2000—a work built up from a recycled trophy kill taxidermied ferret, a rabbit head, a fawn head, pig ears, ferret tails, chicken claws, boar tusks, plastic human mannequin eyes, wax, and shellac—she has said that it is:

> about animal rights … It's not ideological, it's about real life and real death … Almost everyone knows something about the reality of animal suffering. It doesn't really matter if the work is understood with anything other than the heart, I would prefer it to be felt, for the viewer to be vulnerable and open up to compassion.

Compassion is perhaps more readily coaxed from her audience in less brutally hybridised pieces. Quite a number of her reworkings of trophy heads or of complete taxidermic animal bodies over the past decade have been encrusted with jewels, sequins, porcelain flowers, and other such materials that may connote beauty

---

6   Online blog, "Lady Lavona's Cabinet of Curiosities," http://ladylavona.blogspot. com/.

Fig. 1.5   Angela Singer, *Dripsy Dropsy*, 2006. Recycled vintage taxidermy
              rabbit, wax, glass, crystals, wood. 21 x 17 x 17cm. Courtesy of the
              artist.

and elicit sympathy, but in a complex and often troubling manner. Examples include *Garden* (2001–2006), *Deofrith II* (2003), *Recovered* (2004), and *Plume* (2008–2009). She describes *Deofrith II*—the trophy head of a deer covered in old porcelain and bronze grave flowers and flower brooches—as "one of my memorial works. The animal, having no grave site, no bodily burial, becomes its own memorial." The artist turns the already-dead animal (or animal remains, to be more precise) into an object, a different kind of object, and the object then "works," as it were, on the animal's behalf. Animal advocates unsympathetic to contemporary art sometimes criticise artists for "objectifying" animals—Singer herself reports being accused of turning "gallery walls into open graves"[7]—but her work offers one of the clearest examples of the unsettling power of the animal-as-object.

As Singer has recently observed of her working methods, in an interview for the Spanish magazine *Belio*:

> For some artists the material they use isn't important, it's just a way to achieve the object. For me the material, the animal, is everything. Working with the animal body makes me want to investigate what it could have to do with me, with the relationships I have with animals in the world. It confronts me in the safe space of my studio with real everyday brutality ... The animal challenges me to accommodate the frightening.[8]

As that last comment perhaps suggests, Singer is not herself immune to the effects of her chosen working methods. She notes:

> The brutality of deconstructing taxidermy is unavoidable; it jars the senses, it can be quite unsettling to break apart a fragile, vulnerable body. I don't see any way around the process as I need to make the inconspicuous appear conspicuous. Sometimes I think, "I'll write about this instead" but I don't because my artwork can express this further than I can in words.[9]

I am not sure that this is necessarily *always* the case. Describing her early installation *Ghost Sheep* (2001), made of 240 discarded sheep skins from a processing plant, she says:

> I requested one batch of sheep skins to soak, stretch back into their natural form and dry ... The skins are a grey chalky silver blue and have a glow in the dark quality. I hung them high, using invisible thread, near the gallery ceiling to form a tightly grouped ghost flock. The movement of visitors walking under the work made the flock "shiver."

---

[7] Singer quoted in Baker, "'You Kill Things,'" 80.

[8] Singer quoted in Hofstatter, "My Dearest, Dearest Creature."

[9] Ibid.

Fig. 1.6   Angela Singer, *Spartle*, 2009. Recycled vintage taxidermy hawk, modelling clay, wax, wood. 23 x 17 x 17cm. Courtesy of the artist.

In a letter she sent to me at that time, she wrote of having "been to freezing works ... and seen the sheep arrive for slaughter":

> They sense danger from the scent of blood. The stench on the killing floor is revolting ... The sheep and lambs bleat continually, they shake with fear, shitting and pissing *en masse* ... The sheep skins are putrid by the time they reach the pelt processors ... I was not able to think of the skins as anything other than whole living sheep ... Their skins bear the history of their lives and I feel like their witness.

In that instance at least, and even though I've only seen photographs rather than the art installation itself, her graphic description might just surpass the effect of the shimmering skins in the gallery.

The works from a few years later are often more direct, concise, and accomplished. *Dripsy Dropsy* (Figure 1.5) presents the stark image of a recycled taxidermy rabbit head fused with wax, vintage crystals, and red glass beads, the whole thing encased in a vintage glass dome. She offers this terse description of it: "A work with two sides—repulsion and attraction, the beauty of the animal and the ugliness behind its death. Death meets decoration in an undoing of the Victorian diorama. *Dripsy* is not lifelike or posed within a "natural environment," it's in sterile glass, a head on a metal stake." But here it is indeed the image, rather than the description, that is unforgettable.

In relation to what she calls the "flawed dead animal" of her recycled and "botched taxidermy," Singer says: "I don't see an animal separate from myself; there is permeability to the boundaries separating other species from us ... it draws me closer because it's not beautiful, not sentimental, not what animal art is meant to be ...."[10] As she insisted some years ago, art of this kind "should be done strongly, and for me that means using animal bodies that retain the look of a living body because the animal body speaks to the viewer's human body. Lines of body communication are opened up. In our gut we know human and animal are interdependent."

How to make works that address that conviction is an ongoing challenge. Of Singer's newest works at the time of writing, *Spartle* (Figure 1.6) is particularly compelling because its look—somehow arrived at in the coming-together of the recycled taxidermy hawk, modelling clay, and wax of which it is comprised—is utterly baffling. There is something terrible in this object's floundering, flailing, foundering, failing-to-be-an-animal. And its power as an object is in that very instability, that inserting of an instability into human expectations of the natural world. It's a useful reminder that artworks are objects, not ideas, and that they have to work as objects. Whether and how they do so will shape what they can convey or communicate. These things are determined by the *form* of the artwork, by the thickness of its form, and in many cases by its resistance to ready interpretation.

---

[10] Singer quoted in Aloi, "Angela Singer," 13.

In a particularly powerful statement about contemporary animal art in her interview with Giovanni Aloi, Singer says:

> Work that seeks to persuade viewers to take a specific form of action can be quite awful. It can also be sanctimonious and literal. Trying too hard to show the issue you're addressing can lead to dull passionless art of little interest to anyone except those concerned with the same issues. For me the best art is difficult to "read." Returning repeatedly to an artwork that does not give up its meaning easily is a great joy. A great infuriating joy.[11]

***

So, back to the question with which this essay began: What does art *add*? What does art bring to the cause of animal rights that cannot be delivered by the terrible and necessary photographs and film footage of animal abuse that are all-too-familiar to animal advocates?

It *does not* bring answers, or certainties, or "information" in any straightforward sense, though there's certainly an *intention* on the part of these artists to engage with what they see as truth or truths. For each of them it entails provisional decision-making, a preparedness to make changes, and an acknowledgement of the risk of failure, of the audience not getting it, or getting it "wrong." Yet this uncertainty can be seen as something to be welcomed. As Sue Coe notes, "What is intriguing about being an artist, over most other professions, is that there are no right or wrong answers, there is only the search for meaning."[12]

These may not sound like clear positives, so, what else? It brings questioning, and an avoidance of the *easy*. Jaschinski talks of her attempt to "lure the audience into something." Singer says: "I want the audience to come away with questions, not obvious answers."[13] But she has in mind a purposeful questioning: people "understand what an animal is meant to look like. If it has been altered they know it and they can question why."[14] Art also, for these artists, unapologetically brings ambiguity—"the most political art is the art of ambiguity," Sue Coe has said[15]—and at all costs it involves the avoidance of "preachiness." Each of them agrees on this. It involves the attempt to turn meanings around (as in Singer's taxidermic transformations) and a belief in the possibility of using art to see animals differently, to see them anew. But here again this is all about attempts, and beliefs, which may or may not come to anything.

What else does art add? This question may sound vague, but it is important to ask. In the case of all three artists, there is what I can only call a *warmth* to

---

11   Ibid., 17.
12   Anon., "Sue Coe Interview."
13   Singer quoted in Aloi, "Angela Singer," 17.
14   Singer quoted in Baker, "'You Kill Things,'" 93.
15   Coe quoted in Slavick, "Art♢Activism."

the work—in Coe's sympathetic rendering of the humans who get caught up in the industries of animal abuse, in each of the artists' care in the handling of their materials, and, of course, in their concern for the animals that are the subject of their work. In conversation recently, expressing surprise that Wolfe should have raised the "What does art *add*?" question at all, artist Yvette Watt offered a similar suggestion in more articulate terms: "The one thing it really adds is somebody making a considered emotional response."

Along with that considered emotional response, for each of the three artists discussed here, goes a formal rigour; a formal toughness; and an understanding of the medium being used, of the history of that medium, and of the scope for working with *or against* that medium. And for each of them—and this is what separates them (and a few others) from many equally well-intentioned artists with similar convictions—it involves vigilance in never allowing this warmth to slide into a comforting sentimentality. If it is worth doing at all, then to borrow Angela Singer's words, "it should be done strongly."

Does that guarantee that the art will "work," that it will indeed "add" something? No, it does not. But for each of these artists it is a matter of being, and working, in the difficult messy middle of things, and *still* trusting in the processes and the objects with which they work. Let me try to clarify that by contrasting it with a particular philosophical perspective or mindset. In the course of a panel discussion at the 2009 Minding Animals conference in Australia, Dale Jamieson offered the view that philosophy's key contribution to debates about animal rights, and its primary responsibility in that context, was to offer a "consistent" and "coherent" perspective on the issues. This view was not challenged by the other philosophers on the panel, Peter Singer and Bernard Rollin.[16] Whether or not such a view is a sustainable one in a contemporary context is an open question, but it reflects a confidence among some philosophers about philosophy's ability to engage with the world and to achieve its intended effects. The philosopher Cora Diamond— who would certainly distance herself from that confidence—has nevertheless acknowledged in relation to questions of animal life that there is indeed a widespread *expectation* that philosophy (as distinct from novels and poems, for example) is "meant to settle" things.[17]

And one of the ways in which the question "What does art *add*?" might be answered is by saying that artists generally understand and acknowledge something both of the messiness of the world and the messiness of their work, especially in

---

[16] Jamieson, "Does Philosophy Have Anything New to Say About Animals?" (Discussion panel, Minding Animals: International Conference, Newcastle, NSW, Australia, 13–18 July 2009).

[17] Diamond, "Difficulty of Reality," 56. On Diamond's distancing herself from an easy confidence in philosophy's grasp of things, see her observation on the following page that the very "language of philosophical skepticism" creates the serious problem that "[p]hilosophy characteristically misrepresents both our own reality and that of others, in particular those 'others' who are animals," 57.

terms of the precariousness of trying to get the latter to impact on the former in any secure or "consistent" or "coherent" manner. Does this mean that art almost invariably fails? It generally fails "to settle" things, admittedly, but it very seldom sets out to do that in the first place. It would be too easy to say that it more often *unsettles* things, though its effects are often regarded as unsettling, and for all three of the artists considered here there is evidence of their preparedness to embrace the unsettling aspects of their practice, both for themselves and for their audiences. In that sense, Guattari's view of art "as an activity of unframing" does indeed seem pertinent.

Analytical philosophers, in particular, have perhaps been slow to recognise that all this messy undoing may be at the heart of the distinctive kinds of formal rigour and ethical engagement that characterise much contemporary art practice. It was very much in this context, and with an awareness of artists' distinctive ways of doing things, that philosopher Rachel Jones, responding to a question from the audience at the symposium *On Not Knowing: How Artists Think*, stated: "I think that philosophers have more to learn from artists than the other way around."[18]

But it is not a competition, of course. It is about seeing and embracing opportunities. "We are the changes we want to see," Sue Coe has said.[19] Glib as that may sound to some, it chimes with Francisco Varela's conviction, in *Ethical Know-How*, that a particularly persuasive model of "ethical behaviour"—and one that contrasts markedly with "the dominant Western tradition of rational judgment"—involves what he calls "a journey of *experience* and *learning*, not a mere intellectual puzzle that one solves." And this only happens, he suggests, "when the actor *becomes* the action."[20] If, just at the moment, it is artists rather than philosophers who are more inclined to articulate that view through their work, Coe may be right to insist: "There is a gigantic maw out there, starving for pictures, an opportunity for artists who can create their own worlds and share them without waiting for permission" from one source of authority or another.[21] As long as it is not forgotten that art's role, and art's strength, does not generally lie in resolving things, settling things, and then putting them comfortably aside, Coe's may be an encouraging message both for contemporary art and for animal rights.

---

[18] Jones, "Value of Not Knowing."
[19] Coe quoted in Slavick, "Art◇Activism."
[20] Varela, *Ethical Know-How*, 32–4.
[21] Anon., "Sue Coe Interview."

Chapter 2
Marching on Thin Ice:
The Politics of Penguin Films

Elizabeth Leane and Stephanie Pfennigwerth

"Barring a last-minute monkey uprising, 2005 looks set to be the year of the penguin." So observed journalist Tim Dowling in the London *Guardian*,[1] not realising that 2006, and possibly 2007, would also qualify for the title. The early twenty-first century saw the public profile of penguins rise to spectacular levels. Dowling's primary reference point is the documentary *March of the Penguins*—the highest-selling French film and second highest-selling documentary in any language. The film was noted as much for the controversy it generated as its commercial success: the birds at its centre were rapidly co-opted for various political causes and *March of the Penguins* became "an unexpected battle anthem in the culture wars."[2] In the same year, penguins featured in the animation *Madagascar* and their own spin-off, *A Christmas Caper*. Then in 2006 came the blockbuster *Happy Feet*, which told the tale of a tap-dancing Emperor penguin. It earned US$43.2 million at the North American box office in its first weekend (trumping the Bond film *Casino Royale*) and, like *March of the Penguins*, took out an Academy Award.[3] The following year saw two more contributions riding on the coat-tails of these Oscar-winning hits: *Surf's Up*, in which a penguin travels from the Antarctic to the tropics to compete in a surfing competition, and *Farce of the Penguins*, an R-rated straight-to-DVD release which parodies the French documentary.

Why this sudden rise in the popularity of penguins, and what might this popularity mean? The first question is misleading, as penguins' fame has increased gradually throughout the last century, spurred by their appearance in films, zoos, and theme parks.[4] Well before the current craze, Desmond Morris had ranked them among the top ten animals that appeal to humans.[5] Their recent popularity is better seen as a peak in a long-standing trend rather than an indication of any cultural peculiarity of the twenty-first century, although there are some period-specific

---

[1]   Dowling, "My Life as a Penguin."
[2]   J. Miller, "March of the Conservatives."
[3]   Ziffer, "Penguins Earn $55 Million."
[4]   Davis, *Spectacular Nature*, 97.
[5]   Morris quoted in Chester, *World of the Penguin*, 77.

factors worth acknowledging, such as the increase of Antarctic tourism and the pivotal role of the polar regions in environmental debate.

The significance of penguins' popularity is harder to pinpoint, given the extensive symbolic repertoire that has been projected onto them by Western culture throughout the last century. Even a cursory survey of prominent cultural uses of penguins reveals that their symbolic function is characterised by contradiction and incongruity. For the founder of Penguin Books, the birds suggested "a certain dignified flippancy,"[6] a perception probably stemming from their tuxedo-like markings combined with their clumsy gait on ice and land: something formal but fun, classy yet accessible. Batman's evil nemesis the Penguin is an animal-loving "gentleman of crime." In the Wallace and Gromit film *The Wrong Trousers*, penguins are comically sinister. In books such as William Baxter's *People or Penguins: The Case for Optimal Pollution* (1974) and Christine Pierce and Donald VanDeVeer's *People, Penguins, and Plastic Trees: Basic Issues in Environmental Ethics* (1995), they act as synecdoches for pristine nature. Like harp seals and polar bears, they have been adopted by environmental groups as poster children for remote wilderness regions. In greeting cards and cartoons (such as those of Gary Larson) they often symbolise lack of individuality, because many species live in large colonies and their members are, to the average human eye, indistinguishable from each other. *Madagascar* and *A Christmas Caper* draw on the birds' seemingly uniform appearance, portraying zoo penguins as an elite, highly organised military squad. *March of the Penguins* and *Farce of the Penguins* both centre—in very different ways—on penguins' reproductive strategies. *Happy Feet* takes the established absurdity of tuxedoed birds and extends the metaphor to tap-dancing, while incorporating contemporary environmental concerns. *Surf's Up* heightens the incongruity by presenting the bizarre spectacle of cold-climate animals hanging ten in the tropics.[7]

All of these projected meanings rely on penguins' perceived likeness to humans: as flightless birds that stand upright, with arm-like flippers, they are particularly readily anthropomorphised. Ornithologist Bernard Stonehouse notes that penguins are difficult to study "sensibly and objectively" because of the sense that "[c]reatures which look so human must behave like humans."[8] In *A Complete Guide to Antarctic Wildlife* ornithologist Hadoram Shirihai cannot resist describing Emperor penguins' leg feathers as "trousers" and the chicks' plumage as "their so-called 'Biggles' suit of down."[9] At the same time, as fish-like birds confined largely to remote Southern Hemisphere locales, penguins act as exotic others to humanity. This combination of strangeness and familiarity is key to their

---

6   Lane quoted in Lewis, *Penguin Special*, 91.

7   See Stephen Martin's *Penguin* for a detailed cultural history of penguins of various species; in particular, chapter 5 contains useful contextual information about the two films mentioned here.

8   Stonehouse, *Penguins*, 55.

9   Shirihai, *Complete Guide to Wildlife*, 59, 61.

wide appeal and flexible cultural significance. Given that penguins do not come with a pre-packaged set of imposed cultural meanings—or rather, that they come with so many that the "image of the penguin" is over-determined and lacking in cohesion—the significance of each popular penguin text can be discerned only through close analysis.

Of the recent spate of penguin films, the most prominent are *March of the Penguins* and *Happy Feet*. In addition to their focus on Emperor penguins, the films have a number of common features. Both rely on logistically and technically complex production methods, but are traditional in narrative form; both are ostensibly about animals yet are, in different ways, enmeshed in human politics; and both have a political "elephant in the room": climate change. Paradoxically, it was the documentary that caused a political storm, although the director reportedly "didn't want to make a film with an overt message."[10] The animation, by contrast, wore its political message on its sleeve, yet was received largely as pure entertainment. In order to understand these reactions—both to the films and their penguin protagonists—we must examine the way in which each film's form influenced its interpretation.

Family Values and Valued Families: *March of the Penguins*

A central achievement of *March of the Penguins* was the capturing of an event—the hatching of Emperor penguin chicks—which happens only in the depths of the Antarctic winter. Darkness, low temperatures, and frequent blizzards tested both the filmmakers and their equipment.[11] But while the film's production was an adventurous undertaking, its narrative form is quite traditional. Specifically (and despite the fact that it is not a Disney production),[12] *March of the Penguins* follows the blueprint established by early Disney wildlife documentaries—the True-Life Adventure series of the 1940s and 1950s.[13] "The Disney [wildlife] movies always told stories," writes Alexander Wilson in *The Culture of Nature*, "and the stories always began at the beginning—the spring, the dawn." In this model the narrative is organised around the natural rhythms of the wild: the age-old, "eternal" cycle of courtship, mating (although explicit sexual acts are rarely shown), the birth and

---

[10]   Wheeler, "Love in a Cold Climate."

[11]   The filmmakers were based at Dumont d'Urville, a French Antarctic research station located around 800 metres from an Emperor penguin colony. In blizzard conditions it took two hours to travel the distance. The average temperature was minus 30° C; winds reached 160 kph, and during the winter months the men had just two hours of daylight in which to film. They suffered hypothermia, cold burns, and frostbite. Wheeler, "Love in a Cold Climate."

[12]   The original film, *La Marche de l'Empereur*, was produced by French company Bonne Pioche.

[13]   Such films include *Seal Island* (1948), *In Beaver Valley* (1950), *The Living Desert* (1953), *The Vanishing Prairie* (1954), and *White Wilderness* (1958).

rearing of young, the annual migration.[14] Documentaries of this style are heavily edited, with straightforward narratives and closed, happy endings.

As Margaret King explains in her analysis of the genre, Disney's selective perception of animal life exploits humanity's desire to find patterns in the natural world that are similar to its own. By subjectifying animals, the Disney format creates audience identification with animal "stars" and arouses empathy with and affinity for their situations. Audiences are encouraged to relate to nature in human terms and to watch how animals "enjoy" family life, how youngsters grow up, learn their "trade," and gain independence. In relating to animals thus, viewers are encouraged to judge them in terms of such human character attributes as beauty and ugliness, virtue and vice, suffering and reward.[15] Species and moral hierarchies, based on human priorities, are asserted and intertwined: "friendly" or "attractive" animals (that is, those with non-threatening, childlike traits such as round body shape, soft fur, or feathers, or endearingly helpless, "clumsy" movements) are individualised, anthropomorphised, and favoured over "villainous" animals[16] (that is, any unreconstructed species, especially those equipped with talons, fangs, or scales, that eats or otherwise endangers the aforementioned favoured species). This superimposition of human narrative plot structure and social organisation onto animal lives results in "nature, but a very special kind: not an ecosystem, but an *ego-system*—one viewed through a self-referential human lens: anthropomorphized, sentimentalized, and moralized."[17]

In presenting the migration, breeding, and feeding behaviour of the Emperor penguin, *March of the Penguins* follows the Disney pattern, albeit with a twist. Rather than the usual scenario of a female bird incubating her egg in a nest throughout spring, the male Emperor incubates the egg on his own feet throughout the Antarctic winter so that the chick fledges at the height of summer, when food is most plentiful. In accordance with the Disney model,[18] this behaviour is treated anthropomorphically. The original French version had actors voicing the penguins' lines ("J'ai froid!" cries a chick to its parents). The US adaptation of the film replaced this explicitly character-based approach with a narration by Morgan Freeman, but its camera work and voiceovers still encourage enough conflation between penguins and humans to annoy some commentators (for example Kelly). And, as in the Disney model,[19] *March of the Penguins* excludes any visual presentation of humans. The not inconsiderable feats of the filmmakers, along with their physical traces (shadows, equipment, tyre and sledge tracks), are exiled to the final credit sequence; the narrative itself is limited to penguins (and occasionally their predators), who can then stand in for humans without the added complication of actual human presence.

---

[14] Wilson, *Culture of Nature*, 118–19.

[15] King, "The Audience in the Wilderness," 61.

[16] Ibid., 66–7.

[17] Ibid., 62.

[18] Wilson, *Culture of Nature*, 129.

[19] Ibid., 130.

March of the Penguins not only fits the model of the True-Life Adventure documentary, it also shares elements with another, better-known genre: the "Classic Disney" animation, which evolved from the 1930s onward.[20] This is not surprising. Janet Wasko, in *Understanding Disney*, notes the similarities between Disney nature documentaries and animations, pointing to shared predictable plot structures, specific values and ideologies, and formulaic characters.[21] The animations, more specifically, are "typified by light entertainment, punctuated with a good deal of music and humour which revolved mostly around physical gags and slapstick, relying heavily on anthropomorphized (human-like), neotenized (childlike) animal characters." The narrative "almost always" centres around the protagonists' "quest for love," and themes include individualism; escape; innocence and the process of "coming of age"; and the triumph of good over evil.[22] In *March of the Penguins* the birds have a preordained destiny—to perpetuate the species in time for the annual trek to the food-rich waters of the Antarctic summer. Yet this goal is not presented in such bald biological terms. Camera operator Jérôme Maison said: "We weren't talking just about birds when we planned what we were going to film each day, but about characters expressing themselves."[23] Thus, the film is a "love story": its tagline, for example, reads: "In the harshest place on Earth love finds a way." Morgan Freeman, who according to one critic "speaks in the comforting avuncular tone of a wise old fabulist relating a children's bedtime story,"[24] prefaces the birds' initial trek to the breeding colony with the words, "Like most love stories, it begins with an act of utter foolishness."[25] Like the archetypal hero, the penguins must encounter and endure many perils to fulfil their mission, including cold, hunger, and predators, before the goal is reached and the narrative closes on an optimistic note. As in the Classic Disney formula,[26] the heroes of *March of the Penguins* are miniature humans, appealingly childlike while also beautiful in a vaguely aristocratic sense (their plumage provides a suitably rarefied air); the villains are ugly and unambiguously evil (serpentine leopard seals, with gaping jaws; a dark and screeching giant petrel, death on wings—the complexities of the Antarctic food web are not discussed). Comic relief is provided by cute chick sidekicks and bumbling bird-brained adults tripping over ice floes and falling through holes. And even though some of the difficulties of penguin life are not avoided, if only to highlight the birds' eventual triumph against the odds—eggs crack, chicks freeze, predators devour—everything is ultimately entertaining, and there is no chance of an unhappy ending.

---

[20] Early examples include *Snow White* (1937) and *Bambi* (1942).
[21] Wasko, *Understanding Disney*, 146.
[22] Ibid., 111, 115–19.
[23] Maison quoted in Wheeler.
[24] Holden, "A Reprieve for Reality."
[25] Jacquet, *March of the Penguins*.
[26] Wasko, *Understanding Disney*, 114–16.

Given this focus on the anthropomorphically presented "animal family,"[27] it is no surprise that the film was held up by segments of its audience as providing a model of family dynamics to which humans might aspire. Less predictable was the specific symbolic work to which penguin families were put. With their inversion of the stay-at-home mum and breadwinning dad scenario, one would think Emperor penguins might feature in progressive campaigns for paternity leave and hands-on fatherhood. In fact, it was conservative Christian groups in the United States that celebrated the film as a vindication of their "traditional" values. The conservative critic Michael Medved declared *March of the Penguins* "the motion picture this summer that most passionately affirms traditional norms like monogamy, sacrifice and child rearing."[28] Scientists and opponents of the Right pointed out, predictably, that the birds are mostly monogamous for one season only (something the film itself makes clear); that the two parents spend almost no time together, as one has to feed while the other incubates the egg; and that once the chick fledges, they never associate with it again. Nevertheless, the claims for the film's endorsement of "family values" continued. "You have to check out 'March of the Penguins,'" Rich Lowry, editor of the conservative magazine *National Review*, told the National Conservative Student Conference of the Young America's Foundation in August 2005: "... penguins are the really ideal example of monogamy ... the dedication of these birds is just amazing."[29]

The anthropomorphising, Disnified format of *March of the Penguins* is a necessary but not a sufficient condition for its enthusiastic adoption by promoters of traditional family values. To fully understand the phenomenon, it is also necessary to consider the broader socio-political context. Controversy about penguin families had been rumbling in the media in the period prior to the release of *March of the Penguins*. In 2004, Roy and Silo, a pair of "gay" male chinstrap penguins in Central Park Zoo who "adopted" a chick, caused alarm in some quarters and celebration in others (as did their subsequent "break-up" and Silo's new relationship with a female penguin).[30] A children's book based on the "adoption" incident, *And Tango Makes Three* (2005), fuelled the debate, generating complaints from librarians and educators, and topping the American Library Association's 2006 list of "most challenged books."[31] Reports of other same-sex pair bonds between captive penguins in zoos in Edinburgh, Bremerhaven, and Japan suggested that Roy and Silo's partnership was not a one-off occurrence.[32] The controversy degenerated at times into farce, with staff at the Sea World theme park in Brisbane, Australia

---

[27] See Halberstam, "Animating Revolt/Revolting Animation," 267 for a critique of *March of the Penguin*'s insistent emphasis on "repro-heterosexuality."

[28] Michael Medved, quoted in J. Miller, "March of the Conservatives."

[29] Richard Lowry, quoted in J. Miller, "March of the Conservatives."

[30] See Chris, *Watching Wildlife*, 154–5 for an analysis of the media response to Central Park Zoo's "gay penguins" in 2004.

[31] American Library Association, "'Tango Makes Three.'"

[32] Harding, "Females Flown In."

re-badging the species commonly called "fairy" penguins with their official name, "little penguins," due to a misplaced fear of offending the homosexual community.[33] By centring on a heterosexual "love story" achieved seemingly against all odds, *March of the Penguins* offered conservative audiences comfortable certainty in a world of high divorce rates, family breakdowns, and liberal sexual politics.[34]

One feature of *March of the Penguins* which enabled its deployment by Christian conservatives is its presentation of penguin existence as eternal and unchanging. Again, this can be tied to genre. Disney's wildlife documentaries portray "a nature 'in balance' and somehow outside of history,"[35] a "fantasy of pristine nature far removed from the commercial world of modern, industrialized America."[36] According to Wilson, the "cycle of the seasons—'always enthralling, never changing'—sits in for real historic change," and the animal groups depicted can then function as analogues "for conventional ideas about the 'unchanging nature' of human society."[37] *March of the Penguins* follows this pattern: the penguins, viewers are told, have been making their home on the ice for "thousands of generations," but no mention is made of evolution, adaptation, or fossil evidence. Their trek is framed in mythic rather than scientific or historical terms.

A corollary of this ahistorical format is that human interference in the penguins' lifecycle is inevitably ignored. *March of the Penguins* director Luc Jacquet hoped that audiences would be "more concerned about the future of the planet"[38] after watching his film. But the film's adherence to the Disney format works against the interpretation Jacquet sought. Penguins' existence cannot be depicted as both timeless and subject to human interference. Thus, unlike many contemporary wildlife films, which often make at least a token gesture towards environmentalism in the form of a final "message of advocacy for wild animals and their habitats,"[39] *March of the Penguins* makes no reference at all to the environmental hazards facing its subjects, such as the break-up of sea ice, which poses both a direct threat to the penguins' breeding ground, and an indirect threat in its impact upon their food supply. This is a conspicuous absence given the prominence of the climate change debate at the time of the film's release (Al Gore's *An Inconvenient Truth* won the Academy Award for Best Documentary the year after *March of the Penguins* received its Oscar). *March of the Penguins*' implicit depiction of a static ecosystem—a likely factor in its enthusiastic reception by conservative

---

[33]   Tame, "Gay Old Time."

[34]   Conservatives not only embraced *March of the Penguins* as an example of Nature's endorsement of traditional family structures, some also claimed it for a more specifically religious purpose, as "proof" of Intelligent Design. Discussion of this reaction would require another essay.

[35]   Wilson, *Culture of Nature*, 119.

[36]   Mitman, *Reel Nature*, 110.

[37]   Wilson, *Culture of Nature*, 154.

[38]   Jacquet quoted in Wheeler, "Love in a Cold Climate."

[39]   Armbruster, "Creating the World," 218.

American audiences—means that although the film increases viewers' knowledge and appreciation of the species, it provides little motivation for their protection. Thus the political controversy generated by the film, which could have circulated around carbon emissions and melting sea ice, instead focused on matters entirely irrelevant to the welfare of the penguins themselves. Animals became symbolic pawns in human debates, and real penguins disappeared beneath the fuss.

Marching to the Same Beat: *Happy Feet*

While the creators of *Happy Feet* worked with animation rather than live footage, like the documentary makers they went to great lengths, geographically and technologically, to bring detail and accuracy to their representation of penguins and their environment. Two expeditions were sent to the Antarctic, with camera crew capturing the "textures, light and landscapes" of the icescape and its creatures, which were then reproduced digitally through CGI.[40] Considerable attention was paid to the appearance of details such as fur, feathers, and snow under various conditions; the penguin protagonist, Mumble, could boast six million computer-generated feathers. The film's dance sequences were achieved through motion capture, with human dancers going to "penguin school"[41] to learn how to imitate elements of the birds' actual movement.

But despite this drive for verisimilitude, and the fact that, like the US adaptation of *March of the Penguins*, *Happy Feet* was distributed by Warner Bros, at the level of narrative it conforms to the basic conventions of Classic Disney animation. Mumble is a young Emperor penguin who is accidentally dropped onto the snow during his incubation by his distracted father. This early incident means he is unable to sing like his fellows, but has instead a gift for tap-dancing.[42] His colony, led by the rigidly religious Noah (an old, emaciated penguin in true Classic Disney stereotype[43]), believes Mumble's dancing to be the cause of the decline of the penguins' food source. Accompanied by a group of fun-loving sidekicks from another penguin species (Adélies, who, oddly for birds named for a Frenchwoman, are portrayed as Latino),[44] the rejected Mumble leaves his colony, and his childhood love, Gloria, and goes off in search of the real source of the fish shortage and (implicitly) self-affirmation. Following a fishing fleet, he ends up captured in a penguin display in (presumably) a Sea-World style theme park, but through his tap-dancing is able to attract human attention to the plight of his colony. The end of

---

[40]   George Miller quoted in "Production Notes."

[41]   Abbey quoted in J. Miller, "March of the Conservatives."

[42]   Mumble's inability to sing is particularly debilitating because it is through their "heartsongs" that the penguins court each other—an idea based on the scientific observation that Emperor penguins recognize each other through their distinctive calls.

[43]   Wasko, *Understanding Disney*, 116.

[44]   See Halberstam, "Animating Animation," 273, for a brief discussion of the "rather strange ways" in which the film "performs race."

the film sees humans coming together to form policies limiting fishing in Antarctic waters. By this time Mumble is reunited with his family and his colony, his tap-dancing vindicated as an unexpected means of interspecies communication, his future with Gloria now unimpeded by his lack of a singing voice.

The film thus centres on a coming-of-age narrative, with its central moral the expression and acceptance of individuality—both elements of the Classic Disney formula as outlined by Wasko. Since penguins are often considered the epitome of uniformity, using these birds to explore a tale of individual difference generates a nice irony, but it also creates problems, as the main characters need to be distinguishable at a glance by human audiences. Hence Mumble is the only blue-eyed penguin and keeps his fluffy fledgling feathers far longer than his fellow chicks (he is thus doubly neotenised). Other characters are based on well-known human figures with recognisable speech patterns or physical features: Elvis Presley, in the case of Mumble's father Memphis, and Marilyn Monroe, in the case of his breathy, beauty-spotted mother Norma-Jean.

Happy Feet's adherence to the Classic Disney formula means that it shares important features with *March of the Penguins*. Like the documentary, *Happy Feet* is at one level a love story, or rather two love stories: that of Memphis and Norma-Jean, and that of Mumble and Gloria. All partnerships depicted are heterosexual, and monogamy is unquestioned: Memphis and Norma Jean declare their life-long love; Gloria's fidelity is never in doubt; and aural and visual symbols of romantic love—love songs and love hearts—abound. Saddled with a species in which male and female are notoriously hard to tell apart, the animators bestow upon their characters exaggerated gendered characteristics: Memphis towers over his "wife," Norma Jean and Gloria are given hour-glass figures and pink stripes on their beaks, while their golden neck plumage extends further than the males' to provide a subtle suggestion of cleavage. Female penguins feature as mothers, teachers, and love-interests; all questing and adventuring is left to Mumble and his amigos (Gloria's attempts to join them are rejected as putting her in too much danger). Conservative groups, then, could be just as happy with *Happy Feet*'s depiction of "family values" and gender norms as they were with *March of the Penguins*.

Happy Feet does, however, differ ideologically from *March of the Penguins* in an important way. It presents penguin "values" as subject to change, rather than static and eternal. The colony must abandon its rigid, superstitious religious beliefs, and humans must abandon their unthinking practices, before resolution can be achieved. Modern science (absent from *March of the Penguins*) is depicted as saviour: interspecies communication is achieved only after Mumble is equipped with a satellite transmitter and released—a penguin cyborg—into the "wild." The film features striking and powerful images of environmental degradation and human exploitation, including an excavator plunging like a metallic sea-monster into the frozen ocean, a six-pack ring choking a rockhopper penguin, an abandoned whaling station littered with rubbish and animal bones, a fleet of huge, industrial fishing trawlers hauling up nets bulging with fish, and a penguin enclosure whose inhabitants are zombified by their deadening existence. Pollution, over-fishing, and exhibition of Antarctic animals for their entertainment value are all implicitly criticised. The complex realities of actual penguins—subject to human interference, monitoring, and capture—seem ironically more visible in the animation than in the documentary.

However, this environmental message sits at odds with the Classic Disney format; a common criticism of *Happy Feet* by reviewers is its eleventh-hour slip from a film about "diversity and acceptance" to a "screed against human encroachment on [penguins'] habitat."[45] And, as in *March of the Penguins*, a glaring environmental threat to penguins—global warming—is overlooked. The complexities and long time-scales involved in this issue were presumably too difficult to incorporate into a narrative of an individual penguin's coming-of-age; overfishing is a more concrete, less contested, and more readily understood concern. Again, then, the plight of real penguins comes second to the need for a familiar story arc. *Happy Feet*'s glossing of environmental threats to penguins is compromised by a deeper allegiance to the conventions of Classic Disney animation.

A similar argument applies to the film's strategic use of anthropomorphism (which also reflects the influence of the Classic Disney model): while "good" animals, including all of the penguins and several other primarily comic characters (such as elephant seals and skuas), in *Happy Feet* are anthropomorphised, dangerous "bad" characters, including leopard seals and killer whales, are portrayed straightforwardly as animals. As other critics have argued (for example Bekoff),[46] a properly critical anthropomorphism can work positively for animals, disturbing the artificial boundaries erected between humans and other species, and dissolving assumptions that animals cannot experience suffering or pleasure in ways comparable to humans. However, anthropomorphism can also operate in reverse, suggesting that animals are only of value insofar as their differences from humans can be ignored. In *Happy Feet,* the "good" characters are acceptable to the extent that they can mimic human actions. In the dystopic zoo scene, human arrogance appears to be criticised: Mumble, by now well established as the subject of audience sympathy, not only loses his freedom but also, according to the narrator, his mind. He is subjected to the constant yet coolly detached scrutiny of milling zoo visitors, peering through the thick glass of his enclosure. Yet their attention is

---

45   Horwitz, "The Family Filmgoer," and see also, for example, Mayo, "Lovable Happy Feet"; Long, "Happy Feet"; Gillmor, "Happy Feet Awkward." Another problem with the grafted-on environmental message is that the filmmakers' efforts to reproduce in minute detail the physical environment of Antarctica sits in stark contrast to their apparent ignorance of its historical and political context. The concluding images of protests and international meetings creating laws to address overfishing in Antarctic waters blithely ignore the fact that international legal instruments for the protection of the Antarctic environment, and of its marine creatures in particular, already exist: the 1991 Protocol on Environmental Protection to the Antarctic Treaty and the 1982 Convention on the Conservation of Antarctic Marine Living Resources.

46   Bekoff, "Wild Justice," 495.

only truly engaged when Mumble abandons his attempts to shout for their attention (his intelligible pleas sound, to their ears, as unintelligible penguin squawks) and instead executes a soft-shoe shuffle across the artificial snow. In another familiar Disney conceit,[47] it takes an innocent, a child, to truly recognise Mumble's moves and cross the species barrier to mimic them; delighted, she points him out to her hitherto uninterested adult companions, and a media frenzy erupts.

This is the great (and no doubt unintentional) irony of *Happy Feet*: the film emphasises the importance of accepting difference—at the level of the individual (Mumble's eccentricity) and the species—but its success hinges on the flattening of difference. Ostensibly, *Happy Feet* encourages humans to examine their impact on other animals—to see things from another species' point of view—with the concluding section suggesting that humans can better engage with the planet if they adopt a child-like willingness to view all species as equal. However, the humans in the film (and, by extension, its audience) care about the penguins' plight only when they behave like humans—not in the basic sense that they suffer like humans when their food source is depleted, but that they *tap-dance* like humans. The use of CGI technology is telling here: dancers may have been taught to move like penguins, but only so that animated penguins could dance like humans (why not, otherwise, simply create dancing penguins based on penguin physiology?). The irony was not lost on critics: "Thank God those penguins can dance just like humans, eh?" wrote film critic Peter Bradshaw in the *Guardian*, "It means they deserve to live!"[48] Perhaps, for filmmakers, the ability to entertain is as fundamental as the ability to suffer or feel pleasure; but the message is, unavoidably, that penguins deserve to live to the extent that their difference from us is erased.[49]

***

At a time when industrialised human cultures are increasingly divorced from nature and wilderness, and animal life increasingly used to represent human life and values, animal films, both live-action and animated, double as an art form and a social document. Films like *March of the Penguins* and *Happy Feet* have "disproportionate cultural leverage"[50] due to their adoption of Disney's well-tried formulas. But, in both cases, genre imperatives, while instrumental to

---

47   Murphy, "Whole Wide World," 126.

48   Bradshaw, "Review of *Happy Feet*."

49   Sarah E. McFarland's "Dancing Penguins and a Pretentious Raccoon," which appeared after the present essay was submitted for publication, reinforces a number of the observations about *Happy Feet* made here. McFarland labels the film "pseudo-environmentalist," noting the failure to mention global warming and melting sea ice (101–3) and the demand that "nonhuman animals become more like humans before they are recognized" (90). She too points to problems with the film's form, specifically that the happy ending leaves viewers "without a call to action" (95).

50   King, "Audience in Wilderness," 61.

the films' popular appeal, ironically sideline the very animals they purport to portray. *March of the Penguins*, in depicting its subjects in the Disney mode as ahistorical, unchanging creatures isolated from human action, encourages socially conservative interpretations and precludes mention of global warming and its effect on the Antarctic ice. *Happy Feet* presses a last-minute environmental message on its readers, but its Classic Disney emphasis on individualism and coming-of-age, and its homogenising anthropomorphism, undercut its apparently progressive agenda. Each film activates different aspects of penguins' complex and contradictory symbolic repertoire, but the two are united in their inability to go beyond the symbolic animal to address the very real threat that humans pose to Emperor penguins' icy habitat.

Chapter 3
The Traumatic Effort to Understand: Werner Herzog's *Grizzly Man*[1]

Undine Sellbach

I

This chapter explores the relations between human beings and wild animals depicted in Werner Herzog's *Grizzly Man*—a film where the effort to understand these relations is itself presented as traumatic and perplexing.

Grizzly Man is a documentary film that relates the life and horrific death of Timothy Treadwell—a man who longs "to leave the confinements of humanness" and join the bears of the Alaskan wilderness.[2] Treadwell spent 13 summers living in close proximity to grizzly bears both at the Sanctuary (a national parks protected area) and in a less protected area he called the Grizzly Maze. When Treadwell was not with the bears he established himself as a minor celebrity, talking about his experiences to school students and the media, and working as co-founder of the activist organisation Grizzly People. During his last five summers in Alaska, Treadwell took a video camera, and filmed more than 100 hours documenting his life with the bears. In *Grizzly Man*, this footage, edited and cut retrospectively, and combined with interviews from friends, family, colleagues, wildlife experts, and critics, is linked together by Herzog's narration.

Grizzly Man opens up an unlikely conversation between two filmmakers—Treadwell and Herzog—who never met. Treadwell is the film's protagonist. Speaking to his camera in the cute, affected tones of a Disney character, he gives a first-hand account of his life with the wild bears of Alaska: "No other human has survived with the Grizzlies without a gun for this long. The bears will not see my fear—I will hold my ground ... I will die for them, but not at their claws and paws ... They will not kill me because I am their protector and friend."[3] Later in the film Treadwell describes in a letter (read out to the camera by his human friend Gaude),

---

[1] This chapter was developed with support of the Animal Studies Group at the University of Tasmania. I would also like to thank David Wood and Sarah Ernst for their thoughtful comments.

[2] *Grizzly Man*. All quotes are from *Grizzly Man* unless otherwise indicated.

[3] By admitting his fear of being eaten to an imagined human audience, and by hiding this fear from the bears in order to appear as their "protector" and "friend," Treadwell at once evokes a harmonious union with nature, and undercuts this desire. As his words suggest, his safety rests on the bears' inability to read his true feelings.

the experience of dwelling in close proximity to the bears: "I have to mutually mutate into a wild animal to handle the life I live out here."

Herzog's narration carries with it quite different associations: he is the acclaimed European director who has secured the rights to Treadwell's footage. His deep, commanding voice contrasts with Treadwell's rapid squeak. "In all the faces of all the bears [Treadwell] ever filmed," Herzog states, "I see no more than the overwhelming indifference of Nature." There is "no kinship, no understanding … no such thing as a secret world of bears," only "blank stares of indifference and appetite for food." Animals, "in their joys of being, grace and ferociousness," are not the "true subject" of his footage. Rather, Herzog suggests, Treadwell tells us something about ourselves.

This chapter explores the dialogue the film establishes between Herzog and Treadwell. In particular I am interested in the mirror image of human-animal relations Treadwell holds up before us. What is it that he shows us about our nature? Is Herzog right when he insists that Treadwell's footage reflects back nothing but our own humanness?

To understand what is at stake in the dialogue between Herzog and Treadwell it is important to consider not only their competing narratorial claims, but also the compelling and at times deeply discordant collaboration they enter into as filmmakers. The struggles and the shared sympathies of this collaboration are all the more difficult to read given that the dialogue between the two filmmakers is constructed retrospectively, after Treadwell's death.

At the end of the summer of 2003, Treadwell, together with girlfriend Amie Huguenard, was violently torn to pieces and eaten by a grizzly bear. During the attack the camera (with lens cap on) was left running. In the film we watch Herzog wearing earphones, listening to the recovered tape, a document that he finds so horrible, so traumatic that he stops it after only a few minutes, insisting that nobody else should hear it: "the tape must be destroyed."

How should we read this directorial ban? Herzog, one might argue, attempts to preserve the couple's humanity, their dignity as people, by shielding the audience from Treadwell's final act of filmmaking—a sound recording that documents the passage from human being to being meat. If so, an act of effacement lies at the heart of this gesture of human respect.

II

Before responding in more detail to the circumstances of Treadwell's death, I want to turn to the documentary footage that Herzog does allow us to witness. In this footage Treadwell appears a deeply perplexing character. As his biography unfolds we discover that he was a failed actor, short-listed for the role played by Woody Harrelson in the television sitcom *Cheers*. In dramatic tones, Treadwell speaks to the camera of his alienation from society, his breakdown, his collapse into alcoholism and crime. But then he discovered the wilds of Alaska and styled himself as a solitary hero, a caretaker of wild animals, the bears' valiant defender.

Ultimately, Treadwell tells us, it is the bears who are his saviours, because they give his life purpose and meaning.

Treadwell longs to escape the human world, to "mutate into a wild animal," but does not recognise that his desire to escape into a primordial state of nature is itself an invention of Western cultural imagination. Commenting on Treadwell's longing to become bear, Sven Haakanson, the curator of the local Kodiak Indigenous Museum, warns that the line between human and bear should be respected at all times. By spending time with wild animals Treadwell habituates them to human presence, and this puts the bears at risk from people such as hunters. And while one might argue that the dangers of habituation are offset by the "good" work Treadwell does during the winter season, travelling to raise awareness about the plight of grizzly bears, it is also true that Treadwell has became for many a figure of derision, an emblem of the "misguided" environmental movement.[4]

Treadwell's self-portrait as solitary hero is also a construct. As the film unfolds we discover that, although his various girlfriends accompanied him quite frequently on his trips, he meticulously avoided documenting their presence. Huguenard, who died with Treadwell in the bear attack, appears only three times in Treadwell's footage, in each case only for a few seconds.

Furthermore, in the volumes of footage Treadwell takes, he does little in concrete terms to defend the bears. The Sanctuary is already a national park and the "intruders" he films in the Grizzly Maze (a regulated but less protected area) appear to be using cameras, not guns. In spite of the apparent harmlessness of these intruders, Treadwell increasingly comes to view the human world, its signs, language, and symbols, as enigmatic threats. He speaks in dramatic terms about a message he finds scratched on a log by the water's edge. "Hi Timothy, see you next summer!" it says, and near these words there is small pile of rocks and a drawing of a smiley face. The presence of the face in particular, he finds deeply perturbing. A directly threatening message would have been easier to cope with, he tells the camera, but the face looks at him, addresses him without signifying anything determinate.

Treadwell's relation to the human realm is traumatic in a psychic sense for he is riven by the unfathomable desires of others and by the troubling question of his place in the world.[5] For Treadwell the human face is not a mark of welcome

---

[4] In a pragmatic sense, the ethical efficacy of Treadwell's actions is highly ambiguous. Although his sentimental attachment to the bears was used to damage the credibility of the environmental movement, in another sense there is no doubt many audiences were deeply moved by the sentiment he expressed. For a discussion of the capacity of sentimental stories to generate social change see Philip Armstrong's essay "Cetaceans and Sentiment" in this volume.

[5] The word "trauma" first emerged as a description of the shock produced by the tearing of flesh, a wound impacting on the organism as a whole. Under the influence of Freud, the medical connotations of trauma were transposed to the psychic realm. In psychic trauma, a subject or community is riven apart, torn by the shock of stimuli too excessive to metabolise. For an account of the etymology of the word "trauma" and its appropriation by psychoanalysis see Laplanche and Pontalis, *The Language of Psychoanalysis*, 465–9.

or friendship, nor even, as Emmanuel Levinas might say, an enigmatic other that calls him to responsibility.[6] The human face is something he longs to turn from. Treadwell, it seems, is not called to responsibility by the human other, but by a community of wild nature.

III

Writing on *Grizzly Man*, David Lulka points to the contrast between Treadwell's "pre-modern" desire to commune with nature and Herzog's "modernist" need to re-establish a firm distance between "man" and "nature." Lulka is critical of the way Treadwell's anthropomorphic and zoomorphic imaginings tend to warp and homogenise the complex relations and distinctions between human and bear agency. But he raises stronger concerns about "the logic" of Herzog's narration, which, he argues, is even more damaging than Treadwell's misguided efforts to become bear. According to Lulka, Herzog reduces the open adaptability of bear "agency" to a mere "automatic tendency." We hear this in Herzog's narration when, for example, he says that in all the faces of all the bears in Treadwell's footage, he sees nothing but "blank stares of indifference and an appetite for food." Lulka's concern is that this reductive account neglects the complex ways in which bears interact with, and are influenced by, the shifting ecosystem about them. By casting the bear as a hostile predator that usurps human beings' place at the top of the food chain, Herzog reaffirms Western culture's deepest fears of nature. Furthermore, the reductive account of instinct he appeals to—instinct as a blind mechanical impulse to consume—ignores the remarkable adaptability inherent to the instinctual life of bears. So, for example, as scientific research shows, grizzly bears spend very little time hunting humans, their diet is omnivorous (like that of most humans), and shifts with the changing territory and season.[7]

But while Lulka is keen to emphasis the agency of the grizzly bears, he is less open to the subtleties of agency at work in the case of Treadwell. Treadwell, he argues, is simply "a victim of his own video production process, whereby he became enamoured with the image he had created, inadvertently producing a caricature of himself in the midst of describing his own exploits." Herzog, he argues, exploits the apparent "irrationalities" of Treadwell's behaviour in his editing of the film: "whether by intent or folly, Treadwell's identity is caricatured in *Grizzly Man*."[8] This caricature, coupled with Herzog's refusal to let the audience hear the tape of the bear attack, leads Lulka to argue that Herzog discredits Treadwell in an effort to lend authority to his account of the murderous struggle between "man" and "nature." "The proof in Herzog's pudding is Treadwell's death, which enables Herzog to form an inexorable arch that decisively reaffirms the modern stance. In short, *Grizzly Man* is a tale that was done before it was told."[9]

---

6   Levinas, *Totality and Infinity*, 39.
7   Lulka, "Consuming Timothy Treadwell," 70–76, 86.
8   Ibid., 69.
9   Ibid., 72.

While Lulka is right to point out the struggle internal to the dialogue between Herzog and Treadwell—a struggle between competing mythic relations to nature—I have reservations about his claim that Herzog discredits Treadwell, or that his directorial interventions foreclose all chance of genuine collaboration. I suspect that if Herzog had directed a film that cast Treadwell as a respectable (but somewhat misguided) grizzly bear activist, there would have been no dialogue to be had in the film. Treadwell, on my reading, is not an unfortunate victim of the pitfalls of amateur filmmaking—as though the desire to film one's life adventure is an embarrassing impulse disconnected from the ways we really relate to nature. Nor is he the passive victim of Herzog's commentary. Treadwell, one might say, is an animal who demands to be understood. To meet this call for understanding I think it is important that we see Treadwell in all his perplexing detail.

IV

On camera Treadwell bravely exposes himself to the extreme dangers of the wild—like a "muppet" version of Steve Irwin, he suggests that he has a special bond with the bears.[10] In an effort to express this shared community, he gives the bears names such as Mr. Chocolate, Pebbles, Wendy, and Sergeant Brown.[11] He greets them daily, mimicking some of their gestures. He watches in awe as they eat, rest, hunt, graze, fish, fight, and defecate. He goes to sleep at night in his tent with his childhood teddy bear. And when a bears dies (for example, a baby cub is killed by a male bear) Treadwell turns to the camera in horror, unable, in that moment, to accept the harshness of nature.

Sam Egli, helicopter operator and assistant involved in the "clean up" after the bear attack, takes these anthropomorphic and zoomorphic experiments to suggest that Treadwell mistook the bears for "people in bear costumes." Egli states: "The only reason [Treadwell] lasted so long was that the bears probably thought there was something wrong ... like he was mentally retarded." Then one day one bear must have "had enough" or else suddenly realised that he looked "good to eat." Not a physical threat and clearly *mentally deficient*, Egli tells us, Treadwell was tolerated by the bears but inevitably eaten.

If Egli is right in his assessment of the situation, then the bears appear to have boldly inverted a famous argument put forward by moral theorist Peter Singer—that certain animals, the great apes for example, should be accorded

---

[10] Steve Irwin, the "Crocodile Hunter," was also killed by a wild creature (he died not from a crocodile attack but from a stingray). But unlike Treadwell, who set off on his expeditions with a home video camera, Irwin's engagement with wild nature carried the legitimacy of major network coverage.

[11] The word "sentimentality" is often used to denote an expression of feeling in excess of the actual demands of the situation. Treadwell is a perplexing figure precisely because he provokes difficult questions about what counts as an appropriate emotional relation between human and nonhuman animals.

basic interests, such as an interest in living, on the basis that at an intellectual level they resemble mentally disabled human beings.[12] The further question, of whether Treadwell appears to the bears as a mentally defective bear, or a mentally defective human, remains unclear. And perhaps the distinction is unimportant for Egli, because in either case Treadwell is recognised as neither proper animal nor proper man. Although Egli is intent on marking the unbridgeable gap between bear and human, he evokes a certain community of agreement between himself and the bears, namely the agreement that Treadwell is an anomalous figure, to be tolerated then destroyed.

In defence of Treadwell, one might add that although he lavishes the bears with words of adoration, romantically picturing himself as their friend and protector, sometimes to the point of repressing their capacity for aggression, he rarely speaks as though he actually knows what is going on in their heads. Whatever the understanding Treadwell imagines between himself and the grizzlies, it does not amount to an agreement with them regarding the categories "human," and "bear," nor their exceptions.

V

As the film unfolds, it becomes clear that Treadwell's need for human acknowledgement is at least as strong as his desire to be welcomed into the secret world of bears. The footage he takes of his life in the Alaskan wilderness is testimony to this double longing, the desires to "mutate into a wild animal" and to be witnessed in the process. But what is it that Treadwell is performing before the bears? And how does this translate into a performance for us?[13] The acting out of this double longing is tied up with the fact that the sympathy Treadwell proffers to the bears is a gift that they cannot in the full sense receive, given that Treadwell's most elaborate declarations of love are spoken on camera to an audience of imaginary humans. Humankind, at one moment ominous threat, is recast as eager audience.

By refusing to edit out this complex mix of emotions, Herzog brings to the surface the contrary sensations of attraction and disturbance that Treadwell excites in his viewers. By casting his interactions with the bears as a return to a primordial nature, Treadwell projects onto the bears the myth of a world where biological life is not yet perturbed by the traumatic presence of otherness. Although this myth disregards the complex ways that ecosystems negotiate relations between

---

[12]　See Regan and Singer, *Animals Rights and Human Obligations*, 120; and also J. M. Coetzee's discussion of this argument in *The Lives of Animals* (Profile Books), 29–33.

[13]　In an earlier work by Herzog, *The Enigma of Kaspar Hauser*, the hero of the film joins a travelling freak show where he performs side by side with a tumbling bear. Perhaps Treadwell, who is after all a performing animal, might have found some form of allegiance with such a bear.

organism, environment, and world, its prevalence in our culture can make it seem familiar and even appealing to audiences.

But unlike most humans, Treadwell sets out to *realise* the phantasmic return to natural origins in concrete terms. Indeed it is his attempt to *literally become bear* that strikes us as most perturbing, senseless, and dangerous. For in the end the bear does not invite Treadwell into its community, but into its stomach. Wild nature bites back, without conforming to Treadwell's projected ideals or even seeming to recognise the gift of sympathetic understanding offered. By attempting to evade the psychic trauma of being with other humans, Treadwell exposes himself to the very physical trauma of being torn violently apart by a wild animal.

While Herzog's directorial ban shields the audience from the intense trauma of the bear attack, this trauma seems echoed in the perplexing character of Treadwell, and indeed in the name Herzog bestows on him—"Grizzly Man," the bear man, but also the man who is turned to gristle. At the end of the film the audience is left doubly perturbed, confronted both with the almost unimaginable horror of the bear attack, and with the deeply disturbing and at times cringeworthy details of Treadwell's life, his fantasised efforts to become bear. Treadwell has offered us his footage in the hope of being understood, but how can we answer this desire when it seems so delusional?

Treadwell is so compelling, so embarrassing to watch not simply because he is wrong but because he presents, in exaggerated form, that same mix of sentimentality and awe, endearment and horror that characterises many of our daily relations to animals. Think of the fluffy cat, the venomous snake—but also the adored pet with its carnivorous appetite or violent fate, the wild polar bear playing with its young on television, the innocent lamb served up for dinner. Our emotive relations with animals are often disrupted by acts of brutality, our scientific interest in the brute force of nature steeped in sentimentality—and all this from the comfort of our living rooms. Treadwell's mistake, or so it seems, was to live this paradox: to act out these conflicting attitudes, not in the comfort of his home but in the wild, to imagine that that they might form the basis of a genuine encounter between "man" and "nature."

The catch, however, is that these paradoxical emotions are the very tools that Western culture has bestowed on Treadwell to help him through the wilds of Alaska. At a safe distance, human beings find ways to compartmentalise these contradictions, and this helps feed the illusion that the human relationship to nature is balanced and coherent. But it is precisely Treadwell's failure to navigate his way that makes his character so powerfully revealing. Treadwell reflects back to us that same paradoxical mixture of horror, affection, indifference, and sentiment that underpins our own responses to nature.

My proposal is that it is possible to understand, not Treadwell's stance *per se*, but something of his desire to become bear, to leave the human world behind. This effort of understanding seems called for precisely because Treadwell's urge to struggle free of human society is the very thing that binds him more tightly to us, his human audience. One might say that in his attempt to escape the human world Treadwell *materalises* the complex symbols and emotions that make up our cultural imaginings of nature.[14] *Grizzly Man* as a whole might be read as a response to Treadwell's call for understanding. Of course Treadwell is dead and he cannot receive the gift of Herzog's film, anymore than the bears can receive the gift of Treadwell's footage. So what might these un-received gifts of sympathy tell us?

VI

In the sections that follow, I want to look at two attempts at understanding Treadwell put forward by the film. The first is Herzog's narratorial verdict read alongside the coroner's report; the second is the strange and compelling collaboration Herzog and Treadwell enter into as filmmakers.

The coroner, Franc Fallico, is the only person in the film beside Herzog who has heard the tape of the traumatic bear attack. Shielded from the original recording by Herzog's ban, the audience hears of these terrible events second-hand, through Fallico. Significantly, his role as chief dissector does not diminish his humanness—rather he has aura of a kindly priest. His task is to magically turn the human remains—both fleshy and electro-magnetic aural—into the story of Treadwell and Huguenard's final moments.

From the record of "Timothy and Amie's last struggle," Fallico tells us "we discover what separates them from wild animals." Amie, Treadwell's diaries state, has been frightened of the bears all along. Yet when Timothy is attacked, she hits the bear with a saucepan and tries to drag the bear off. Timothy begs Amie to run away while she can but Amie refuses to leave. At the final moment, he tells us, they each try to put another life, a *human* life, before their own.

If we put the coroner's report alongside the "blank stares of indifference and appetite for food" that Herzog finds in the faces of the bears, then the film might be read as an illustration of an idea put forward by Levinas—that we are called into responsibility by the enigmatic face of the other. This face is always human, Levinas tells us, for an animal's struggle for survival *overrides its having a face*.[15] In people, on the other hand, there is something more important than the struggle for life, namely the life of another. And here we must think of Huguenard, who quite literally remains faceless in Treadwell's footage until just a short time before her attack when she looks briefly towards the camera. This is the face that Treadwell denied in order to fashion his image of solitary hero, but it is also the face that seems to call him into responsibility at the very last moments before his death.[16]

---

[14]   As Cora Diamond has argued, in some moral contexts the type of understanding called for is not comprehension of the content of the other's claim, but rather an acknowledgement of their predicament. This may at times entail a willingness to imaginatively take nonsense for sense. See her "Ethics, Imagination and the Method of Wittgenstein's *Tractatus*," 61.

[15]   See Levinas, *Difficult Freedom*, 151 and "The Paradox of Morality," 169.

[16]   The singular but enigmatic presence of Amie's face stands in contrast to the generic smiley face left for Treadwell by the tourists. Both induce trauma but, if the coroner is right, Amie's face returns Treadwell to responsibility, whereas the smiley face becomes a pretext for Treadwell's fantasy of escape into nature.

The nonhuman animal, in Levinas's account, is no more than a "hungry stomach" "without ears"; it eats without hearing the other's demand for ethical recognition. Of course, as Levinas makes very clear, human beings can also be "deaf" "stomachs" both literally and metaphorically.[17] However, there is the potential amongst humans to respond to the violence of consumption, with a different kind of relation—a proximity, which is not assimilation; a distance, which is not effacement.[18]

VII

As compelling as this reading may be, I want to raise some points of difficulty. To begin with it is not clear that the digestive workings of the mouth and gut—in human and nonhuman animals—are analogous to the mode of consumption Levinas critiques, a consumption based on the double gesture of assimilation and obliteration. In her extraordinary account of surviving a crocodile attack, eco-feminist Val Plumwood describes the first moment when she truly experienced herself as prey: "In that flash, I glimpsed the world for the first time 'from the outside,' as a world no longer my own, an unrecognisable bleak landscape composed of raw necessity, indifferent to my life or death."[19] Caught in a death roll, Plumwood experiences the crocodile's terrifying indifference to her story. In a real sense, the crocodile is, at this moment, a stomach without ears, a predator unmoved by the singularity of her life. But the resemblance between Plumwood's experience and the ethical deafness that concerns Levinas is partial at most. For Levinas's analysis focuses on a form of consumption that simultaneously claims mastery or knowledge of another. The crocodile is indifferent to Plumwood's life story, but it is equally indifferent to the task of masterful knowing.

The impulse to know at the very point of incorporation and obliteration is exemplified in the film by the fate of bear 141, the grizzly who is discovered not far from the campsite with the remains of Treadwell and Huguenard in his stomach. Herzog refers to this creature as bear 141 because of the identification tag tattooed onto its hide. This brand is a reminder that "brutal nature" is also managed, catalogued by national parks, and finally, in the case of bear 141, shot

---

[17] Levinas, *Totality and Infinity*, 134.

[18] In spite of his desire to distinguish human and nonhuman life (and hence prevent the human from being reduced to the merely biological and the animal sentimentalised as the human) Levinas is sometimes troubled by the strange claims made on us by other animals, by the ethical space of relation that might open up between beings. See his essay "The Name of a Dog, or Natural Rights," *Difficult Freedom*, 151–3 and also David Clark's essay on the ambiguities of Levinas's relation to animals in "On Being 'The Last Kantian in Nazi Germany.'"

[19] Plumwood, "Prey to a Crocodile," 1–2.

and dissected.[20] Bear 141 suffers trauma in the literal sense of the term, that is, his body undergoes a surgical procedure that violently disrupts the organism as a whole. The traumatic death and dissection of bear 141 is an act undertaken in the service of human knowledge. This act enables Huguenard's and Treadwell's body parts to be recouped from the bear's digestive system, and this in turn helps the coroner build a narrative to "make sense" of their traumatic deaths.

What the autopsy disrupts is precisely the form of consumption which fascinated and terrified Treadwell most of all, the digestion of food by the gut and the absorption of bodily waste back into the earth. Treadwell recognises this dimension of consumption when he speaks with reverence to a pile of warm bear scat that has recently been inside one of his favourite bears. But at the same time he separates himself from the ecosystem about him by proclaiming himself as friend and protector of the bears, a claim that expresses his sense of affinity and covers his terrible fear of being absorbed and excreted.

VIII

I want to explore an alternative understanding of Treadwell's efforts to become bear. Alongside the dramatic acting out of human sentimentality and horror that Treadwell's character reflects back to us, or perhaps almost inseparable from this parody, is the fact that Treadwell is not eaten immediately. Instead he and his various human visitors live for thirteen summers in close proximity to the bears. And as the footage shows, between bear and human there are daily encounters, familiarity with some of the other's behavioural traits, mutual negotiation of space, co-habitation, and a certain habituation—almost as if myth, in the process of being acted out, cannot help but begin to unravel.

In these meetings Treadwell's professed bravado translates into cautiously standing his ground, or carefully retreating, not dominating, but trying not to let the bear see his fear. Sometimes the audience sees Treadwell—arms outstretched and hands bent over like paws—approaching the bears; at other times only the bears are in shot, but always the audience hears Treadwell negotiating the situation in his little cartoonish voice. From the footage it is clear that although Treadwell longs to enter into the bear world, to become a bear, during his actual encounters with bears it is often he who draws the boundaries. When, for example, a particularly curious bear approaches, Treadwell tells her in cautious tones that although he loves her she must return to her friends. There is a gap between what Treadwell pronounces to the camera and what he actually does, between the fantasy and its acting out.

---

[20] As Giorgio Agamben points out in *Homo Sacer*, we live in a time when the distinction between the natural and political realms has broken down such that biological life itself is exposed to intensified management. This may mean that we can no longer be sure that nonhuman animals are indifferent to the potentially traumatic effects of this management.

This unsettling mix of closeness and distance becomes manifest in various ways. For example, the very limited encounters negotiated between Treadwell and the bears contrast with the much more complex, interactive engagements that develop between Treadwell and the local foxes who investigate his campsite. Even so, between the foxes and Treadwell there is both nearness and distance.

There are further permutations of this mode of encounter in Herzog's human interviews. They emerge at certain moments of conversation, but often most poignantly when the interview is done and the camera is left rolling. When the national park officials go to the site of the bear attack, they find Timothy's severed arm with the wristwatch still on. Herzog films the scene when the coroner Franc Fallico hands the watch (still running) to Jewel Palovack, who has inherited most of Timothy's belongings. When the ceremony is over, the expected mix of solemnity, sadness, humour, and respect give way to a relation far more difficult to read, an awkward juxtaposition of coroner and "widow."

Indeed it is Treadwell's readiness to document similar unexpected moments that Herzog most admires about his filmmaking. "Sometimes, images themselves develop their own life, their own mysterious stardom," Herzog tells us. There is a "strange secret beauty," to many of Treadwell's shots of landscape. On one occasion, when Treadwell finishes his spiel in front of a grizzly bear grazing in a meadow, two foxes wander in and out of shot. Between foxes, bear, and man there is at once co-habitation, a co-existence, and a remarkable alterity.

These are the scenes where Treadwell and Herzog seem to come closest as filmmakers. At these moments of collaboration, that strange mix of proximity and distance that Levinas describes as the space of human ethical relation is refigured as an encounter between beings, human and otherwise.

IX

Treadwell, I have argued, is so difficult and compelling to watch because he holds up a mirror to the contrary myths and emotions that comprise the Western cultural encounter with the wild. While most people simply compartmentalise these contradictions, Treadwell set out to enact them. The film is disturbing because it makes us aware of the contradictions we navigate in our lives with other animals. In this regard, Herzog is right to point out that it is not *wild* nature but *our* nature that Treadwell's footage reflects in exaggerated form. But caught in this reflection, almost as if by accident, is our awkward proximity to other animals. It is difficult to give expression to the proximity, for it is so closely intertwined with sentimental anthropomorphisms. But it is also difficult to express the *otherness* of nonhuman animals, because these efforts can turn into reductive claims about the categories "human" and "animal." In the effort to articulate this relation, the proximity threatens to collapse into an assimilating projection, and the distance into an assertion about the very nature of the other. Perhaps this marks the need to recognise as unresolvable the question of what, if anything, the bears want from Treadwell, beyond the satisfaction of their nutritional and territorial needs?

Treadwell's longing to be understood, on the other hand, is quite unmistakable. It is unclear, however, whether his efforts to become bear were, in any sense, "gifts" the Grizzlies could "receive." Put in simple terms: a captive audience is hard to find in the wild. For this reason Treadwell's display for the bears is also a display before the camera. It is unlikely, however, that Herzog's presentation of this footage would have satisfied Treadwell. If *Grizzly Man* is a response to his need for understanding, then it does not offer the comfort of making his life legible. Instead the film reveals, in Treadwell's efforts to become bear, the disorientation and torsions of a performing animal.

Chapter 4
Naming and the Unspeakable: Representations of Animal Deaths in Some Recent South African Print Media[1]

Wendy Woodward

Animals appear constantly in South African print media. Not only are syndicated international articles on animal ethology or animal behaviour published, but local and national narratives, which are the subject of this essay, are also featured. National narratives tend to relate to South Africa's position in connection with CITES (Convention for International Trade in Endangered Species), the ignominious, ineffective legislation of canned hunting, the culling of elephants, and the recent inception of elephant training for tourist safaris. Narratives local to Cape Town tend to focus on the horrors of illegal dog-fighting, pit-bull terriers turning savage, or the ill-treatment of animals in the townships, as well as on baboons making apparently military-style forays into human habitation. Animals are differentiated into the category of the "wild," who have to be managed in reserves and along the borders of human habitation, and that of the domestic or farm animals, whose treatment is implicitly and sometimes explicitly racialised in relation to their owners.

What interest me in this essay are the narratives of particular, individual animals, for such stories are far more effective in engaging the reading public than more general reports of animals as species. It follows that the analysis of such material derives from a human-animal studies approach, which takes seriously the issue of the animal as an individual and as a subject. An approach foregrounding species rights would not elucidate the material so productively.

Many representations of animals deal with their deaths or the likelihood of their deaths, and reveal, or occasionally dispute, broadly held notions of the place of animals in modernity. The Biblical injunction of "Thou shalt not kill," which is always read as referring only to humans,[2] is extended in the *Dhammapada,* the classic Buddhist scripture, in which neither those who kill nor those who cause others to kill are "holy." As Jacques Derrida suggests, however, the status

---

[1]   I am very grateful to Marlei Martin, then at the Canned Hunting Campaign, now at Baboon Matters, for access to her press cuttings. I am also indebted to the University of the Western Cape for funding my project on Animals and Cultural Discourses and to Elvera Boonzaier for her committed research assistance.

[2]   Derrida, "Eating Well," 112–14.

of the [human] subject is ascertained by the sacrifice of the [animal] other in a "phallogocentric structure" which "implies carnivorous virility." Thus if we ask, "Do we [in the West] not have a responsibility toward the living in general?" then "the answer must necessarily be 'no' according to the whole canonised or hegemonic discourse of Western metaphysics or religions."[3] But what Derrida terms the "sacrificial structure" of such discourses may feature, too, in some traditional African religions and practices.

The essay will focus on media representations of the deaths of animals who are individualised through naming or specificity—Yengeni's sacrificial bull, the individualised ritually slaughtered chicken at a theatre performance, David the baboon, and Frida the lion cub, all of whom are ultimately depicted as victims. The naming of these animals does not necessarily incorporate taming. While Frida may be a captive lion, David remained a free-ranging baboon for most of his life. In the case of Frida and David, both members of endangered species, at least outside of canned hunting farms, zoos, or laboratories, the media emphasis fell on the drama—and tragedy—of their respective narratives. In the case of the bull that Yengeni slaughtered, who occasioned an extraordinary ongoing media debate, public outrage was directed not at the death of an endangered species but at the treatment of a member of a domestic, flourishing species. Likewise, the extent to which these particular condemned animals are accorded subjectivity, even while they may be individualised, varies. Media representations of animals, at least in my reading of local Cape Town newspapers as well as national weekend newspapers over the last few years, are paradoxical. In connection with domestic and farm animals, an article may seem to rescue an animal or animals from invisibility or anonymity, often quite literally with a photograph, but the representation of the individualised animal, I would argue, renders the animal exceptional in its his/her intelligence, sensitivity, or suffering. When a "wild" animal like a baboon or a lion features in media narratives, on the other hand, the public's eating habits are in no way challenged. Consequently, the stories of David and Frida suggest that their individualities can come to stand for those of other members of their at-risk species.

Such media representations may seem to be offering alternative glimpses into what modernity does not acknowledge about the ongoing killing of animals, but the Derridean "sacrificial structure" is evident in journalistic discourses, with animals as instrumentalised others whose suffering and erasure is rendered unspeakable—and thereby repressed. Many meat-eaters prefer to override the knowledge that an animal has suffered and died for their plates. The conventional marketing of what J. M. Coetzee calls the "muscle flesh of an animal slaughtered by someone else"[4] skilfully attempts to perpetuate the illusion, with animals butchered out of sight and parts of their bodies commodified in pristine packages. Yet such practices may

---

3   Ibid., 112, 113.

4   Coetzee, "Comments on Paola Cavalieri," 86.

exemplify "the perverse human tendency to attempt to deaden self-disgust in ways that actually intensify it."[5]

Richard W. Bulliet usefully differentiates between postdomestic society, which prefers to abolish the staged spectacle of animal killing, and domestic society, in which humans and animals live in close proximity, with the slaughtering of animals an accepted part of everyday life.[6] Within South Africa, both the domestic and postdomestic modes of responding to animals are evident and are generally aligned with racialised differences. The still mostly white middle class is predominantly postdomestic, whereas pervasive traditional African relationships with nonhuman animals tend to be domestic. My drive to the campus of the University of the Western Cape (a historically black institution) dramatises such differences as it exemplifies apartheid geography. After the green suburbs of Cape Town close to the mountain where domestic dogs and cats live invisible lives behind high walls and fences, I pass through coloured townships,[7] then alongside informal settlements which are populated mostly by African inhabitants, some of whom have come to the city from the rural areas. A herd of goats scavenges confidently amidst detritus on the narrow side of a six-lane road; carthorses and donkeys graze, mostly untethered; a clump of pigs occasionally parades close to the busy intersection under a massive billboard; and, slightly removed from the traffic, a herd of cows ranges on the winter grass.

Yengeni's Bull

Such cultural differences in relation to nonhuman animals were dramatised in media representations which made much of exposing the fault lines between tradition and modernity, between sacrificial slaughter and factory farming. As the Animal Studies Group reminds us, "[k]illing an animal is rarely simply a matter of animal death. It is surrounded by a host of attitudes, ideas, perceptions, and assumptions."[8] When Tony Yengeni was released from prison four months into a four-year prison sentence for corruption and lying to parliament, of which he was then a member,[9] he celebrated by slaughtering a bull to the ancestors in gratitude. He had purchased the bull from a farm 200 kilometres away and driven the animal to his parents' home in Guguletu in greater Cape Town in an open truck.[10] The bull, with a red and white brindled coat, was exceptionally photogenic, even as

---

5   Mulhall, *Wounded Animal*, 75.

6   Bulliet, *Hunters, Herders and Hamburgers*, 3.

7   Townships were created by the Group Areas Act (1950, 1966) which master-minded apartheid geographies. All people of colour inhabiting inner cities and suburbs were forcibly removed from their homes and re-settled separately in their designated groups in outlying areas. Townships are thus literal places but also discursive constructions.

8   Animal Studies Group, "Introduction," 4.

9   Burbidge, "The Big Beef."

10   Huisman et al., "Yengeni Cleansing Party."

he was pictured roped and trussed on the ground with his eyes rolling in terror. The dramatic photograph by Fanie Jason appeared on the front page of the *Weekend Argus* which broke the story,[11] and its affective qualities were powerfully instrumental in initiating the enraged debate which followed.

A great deal of readers' ire was raised by the actual mode of killing the bull. Traditionally, a bull is prodded with a spear; if he does not respond by bellowing, his sacrifice is judged not to be pleasing to the ancestors and another animal has to be selected. The representation of the ritual sacrifice exposed to the public gaze the usually unspeakable practice of killing an animal ultimately for his meat—but the differences between the ways of domesticity versus postdomesticity were acrimoniously racialised. Those who were critical of Yengeni were labelled as reactionary whites without awareness or appreciation of Xhosa tradition and black identity enshrined in the Constitution. In the opinion of Sandile Memela, from the Arts and Culture Ministry, cruelty to animals was not at issue: "Instead it is about man's search for meaning, purpose and the redefinition of the relationship with the cosmos, God and his ancestry."[12] Furthermore, Fikile-Ntsikelelo Moya claimed: "In its dedication to the cow's [sic] cause, [the SPCA] betrayed a glaring insensitivity to South African and African history. Its objections came over as a knee-jerk reaction, inspired by the colonial desire to educate the brutish natives."[13] Representatives from the SPCA were quoted at various moments as they investigated charges against Yengeni under the Animals Protection Act. But rather than condemning the practice of animal killing *per se*, they were preoccupied with whether the bull had suffered before the actual slaughter and to what extent his "dignity" had been undermined.[14]

Middle-class readers were offended by the public statement of the unrepentant Yengeni celebrating his freedom, but also by the spectacle of a beautiful, condemned animal who had been controlled with some difficulty, judging by the number of ropes used to subdue him and render him powerless. The photograph also deftly foregrounds the usually unacknowledged hyper-masculinity and violence of killing an animal for meat, with Yengeni in an aggressive pose above the prone bull. The killing of animals for middle-class consumption takes place out of sight in circumstances which are perceived as clinical, orderly, and emotionless on the part of the killers.[15] But such customary "*cold* deaths" achieved mechanistically[16] are contradicted by the photograph of the wild-eyed bull which gives the lie not only to the myth of the absolute control of the defenceless animal about to be slaughtered, but also to professionals remaining dispassionate and emotionally uninvolved in

---

11   Gophe, "Yengeni's Spear."
12   Quoted in Williams and Prince, "Tony Slaughter Critics Lashed."
13   Moya, "SPCA Needs to Work."
14   Williams and Prince, "Tony Slaughter Critics Lashed."
15   Marvin, "Wild Killings," 16–17.
16   Ibid., 17.

Fig. 4.1   The unspeakable pictured: the unnamed but individualised bull in the Yengenis' yard moments before he was sacrificed. Photo: Fanie Jason.

"non-individualized relationship" with the animals.[17] Ritual sacrifice is disorderly and dependent on negative human emotions and the desire for the animal's death, after which there is celebration.

If the media debate was sparked by the specificity of the brindled bull photographed in his last moments, it was motivated and extended by the sentimentality of viewers and readers who focused on this particular animal, often in ways that were blind to their own cultural practices. For Derrida, compassion may be awakened at the spectacle of "*this* irreplaceable living being,"[18] but compassion, at least in the broad Buddhist sense, extends to all beings and recognises their sentience. Sentimentality, on the other hand, sensationalises feelings of pity in ways which are patronising, often foreclosing any acknowledgement of the animal's subjectivity. Instead, the viewer/reader often foregrounds his/her own outrage so that the victimised animal is othered and distanced. Responses to the bull were sentimental as well as hypocritical, for they focused on the individualised bull rather

---

[17]   Ibid.
[18]   Derrida, "Animal Therefore I Am," 378–9.

than on the ongoing practices of slaughter which modernity (or postdomesticity according to Bulliet) represses.

In *The Lives of Animals*, Coetzee implicitly underscores the hypocrisy of this position when he has Wunderlich say:

> The Greeks had a feeling there was something wrong in slaughter, but thought they could make up for that by ritualizing it. They made a sacrificial offering, gave a percentage to the gods, hoping thereby to keep the rest ... Ask for the blessing of the gods on the flesh you are about to eat, ask them to declare it clean.[19]

Such spiritual displacement of the desire to consume meat on to metaphysical approval appears in traditional African notions of sacrificing to the ancestors, an issue which was obvious to most vociferous critics of the practice, but the Judaeo-Christian religious justification for eating nonhuman animals was barely reflected on. An editorial in the *Cape Times* did, unusually, berate "the cultural intolerance of Yengeni's non-vegan critics": "Here we have a blithe view that a religious sacrifice is somehow barbaric and uncivilized when compared to Western factory farming and killing methods. What nonsense. The ritual has significant spiritual and cultural meaning; modern husbandry and retailing is simply about profit."[20] The editorial disingenuously occludes the significance of meat eating in modern cultures, thereby ignoring the "spiritual and cultural meaning" of the widely adopted, but initially Afrikaner, South African ritual of "braaivleis" (barbecuing).[21]

It would be simplistic, perhaps, to speculate about how many of Yengeni's critics became compassionately inspired vegetarians after the spectacle of this bull terrified at his impending death. The *Weekend Argus* of January 27 ran an article on the slaughter of cattle at the Grabouw Abattoir[22] near Cape Town. Generally, however, the debate was marked by silence on the unspeakable daily condemnation to death of hundreds of thousands of animals in Cape Town alone—in so-called meat-processing plants and in informal or illegal slaughtering. The discourse of most articles also disavowed the living animals, rendering them "absent referents," to borrow a term from Carol J. Adams.[23] The belaboured puns of Burbidge's headline "The Big Beef" or of Gophe's "Abattoir Workers Have No Beef with Tradition" depict the animals lifelessly as always already meat and without subjectivity.

---

[19] Coetzee, *Lives of Animals* (Princeton UP), 40–41.

[20] "Cultural Conflict."

[21] Nobel laureate Bishop Tutu proposed that all South Africans should be united in braaing on Heritage Day on 24 September, 2008. A poster depicting a sizzling steak in the shape of Africa was part of the campaign.

[22] Gophe, "Abbatoir Workers."

[23] Adams, *Neither Man nor Beast*, 17.

The SPCA case against Yengeni was finally dropped due to a "lack of evidence of animal cruelty" and "no witnesses."[24] Subsequently, SPCA chief executive, Marcelle Meredith derided the Cape of Good Hope Branch for being "foolish to get embroiled in a political matter."[25] *The Sunday Independent*, which carries balanced articles on the politics of animals' lives, noted in an editorial that no animal rights activist was given an opportunity to engage in debate in order to put forward the position that animals should not be slaughtered at all. Instead, the animal welfare position held by the SPCA which attempts to ensure humane treatment for animals, including those about to be slaughtered, was foregrounded.[26]

Although opportunities were lost for larger debates about the killing of farm animals, some positive outcomes emanated from the Yengeni furore. Mongezi Guma, chairperson of the Cultural and Linguistic Rights Commission, felt that the media interest in the Yengeni case had led to more understanding between "different cultures" and promised to consider how "to do cultural slaughtering in a way that will promote and protect the welfare of animals."[27] While an animal rights position will see this as a contradiction, an earlier article underscores not only the porosity of culture but the possibility that traditional attitudes to animals have been contaminated by modernity's instrumentalising of them as resources. Chief Nyembe, Humane Education specialist and head of *Animal Voice*'s Khayelitsha branch, derided the acceptance of "Western values," which includes the unthinking consumption of animals in large quantities, with the result that animals are treated in working-class areas "as if they have no value of their own."[28]

The Chicken in *iMumbo Jumbo*

When a chicken was slaughtered ritually on the stage of the local Baxter theatre on the final night of *iMumbo Jumbo*, a play by Brett Bailey which "illuminat[es] the tensions between African and Western beliefs and values"[29] and which incorporated *sangomas* (traditional shamanist healers) as actors, many of the same issues connected to the slaughtering of the bull by Yengeni were raised in the press: for example, the imperative of African traditional sacrifice and the racism of its critics. Yunus Kemp's article similarly deployed reductive attempts at humour about animals through puns: "Murder Most Fowl: Theatre Fans Clucking Mad after Baxter Bloodbath." Another mockingly entitled article, "Luck for Plucky Clucker," told how the hen called Veronica who had previously acted throughout the run was spared on the last night and a "stunt double" purchased on the side of

---

24   Peters, "SPCA Finds No Evidence."

25   Sapa, "Yengeni Row."

26   "Facts Distorted in Drama."

27   Sapa, "Yengeni Row."

28   "Khayelitsha Commemorates," 3.

29   Kemp, "Murder Most Fowl."

the road was brought in to be slaughtered.[30] In this way, the actors, who had named Veronica and admired her because she "effectively acted her way out of the death sentence,"[31] demonstrated the dualistic ethics of favouring a named chicken at the expense of an unnamed bird.

Without individuality or subjectivity conferred on her by humans, the latter did not merit what Martha Nussbaum calls "direct obligations of justice" to nonhuman animals whom we need to recognise, ethically, as "subjects and agents."[32] Derrida is very clear about the killing of animals being tantamount to murder, asking rhetorically: "Do we agree to presume that every murder, every transgression of the commandment 'Thou shalt not kill' concerns only man ... and that in sum there are only crimes 'against humanity?'"[33] And J. M. Coetzee has Elizabeth Costello turn to the significantly named Dean Arendt and say: "'What is so special about the form of consciousness we recognize [in humans] that makes killing a bearer of it a crime while killing an animal goes unpunished?'"[34]

The dominant practices of instrumentalising animals and killing them remains, for the most part, unspeakable in the media reports discussed above. Named or individualised, like Yengeni's bull or the *iMumbo Jumbo* chicken, the animals are sentimentalised and rendered exceptional. The bull and the chicken, rather than metonymising others who suffer similar experiences, are represented as distinctive, so that others in their predicaments are erased. Consequently, representations of the deaths of the sacrificial bull and the ritually killed chicken cannot really be effective in deflecting newspaper readers from eating dead animals. Such representations never actually interrogate the practices of modernity which relies on the sacrifice of animals for the constitution of the human subject. The unspeakable practices of slaughtering cattle and chickens on a monumental daily basis are, at least in the initial stories, if not entirely in the ensuing debates, repressed and silenced.

David and Frida

On the other hand, the tropes differ in representations of "wild" animals. Given the dominant view that these animals should be preserved, admittedly for their commodification as tourist spectacle rather than for their intrinsic right to exist, they are not depicted as dispensable like domestic or farm animals. Bulliet confirms that postdomesticity accords "wild" animals the right to life,[35] and while this broad perception disregards specific practices like hunting, it is extant in the cases of David the baboon and Frida the lion cub. When a baboon in greater Cape Town

---

[30] "Luck for Plucky Clucker."

[31] Ibid.

[32] Nussbaum, *Frontiers of Justice*, 51.

[33] Derrida, "Animal Therefore I Am," 416.

[34] Coetzee, *Lives of Animals* (Princeton UP), 44.

[35] Bulliet, *Hunters, Herders and Hamburgers*, 20.

and a lion cub kept on a canned hunting reserve are individualised and named, they do function as metonymic of others of their species, whose plights the reader is drawn into sympathising with. They might be represented as individuals, but they are not rendered exceptional like the bull and the chicken, who are sentimentalised even while the killing of farm animals for food continues apace. In a modernity in which the putting to death of domestic animals is conventionally not constructed as a crime, some of the media articles in relation to the baboon and the young lion do depict animals as worthy subjects of justice.

Baboons are both loved and reviled—and abused, as they are in ongoing scientific experimentation. Animal testing emerged in the Truth and Reconciliation Commission[36] because baboons were used in trials to establish the most effective "lethal biological and chemical agents against perceived enemies of the apartheid state."[37] In rural areas and peri-urban areas where they come into contact with humans they are either celebrated, often for their apparent comic potential, or seen as a dangerous nuisance. In the Cape Peninsula (where I am located), most interactions between resident baboon troops and people in housing developments which encroach on their habitats have been disastrous. South African legislation categorises baboons as "vermin" in spite of being listed by CITES as "vulnerable and only one step away from being endangered."[38] John Yeld, for example, the Environmental and Science Writer for the *Cape Argus*, reports on "the death of five baboons in ten days in Scarborough—four by shooting and one from electrocution in a fight with another baboon while being chased by the baboon monitors who were attempting to keep their charges out of the settlement."[39]

In an expansive article in the *Sunday Times*, entitled "Primate Scream" and under the heading of "Space Invaders," Cosmo Duff Gordon initially conveys Scarborough residents' negative perceptions of the local baboon troop. He writes, ironically, of the way the "dozy tranquillity" of the upper middle-class seaside village "[in] a sort of Stepford Wives way ... hides a heart of darkness, certainly where baboons are concerned."[40] The "heart of darkness" metaphor echoes, possibly, the racist/speciesist homology of baboons with African people in apartheid discourse, also emphasising the danger inherent in interactions with baboons. Gordon refers to the "urban myths" centred on the "superstar" alpha male of the troop, William the Conqueror, who "oozes testosterone and is indeed cleverer than a monkey" (precisely what he means by this throw-away line is unclear). William's masculinity is proven by his exploits characterised as those

---

[36]   The TRC was set up after the first democratic elections held in South Africa in 1994. It recorded testimonies of survivors of human rights abuses as well as those of relatives of murder victims. Perpetrators of violence (from all sides) could request amnesty from prosecution.

[37]   Pickover, *Animal Rights in South Africa*, 130.

[38]   Van Riel, *Life with Darwin*, 175.

[39]   Yeld, "Fears for Baboons."

[40]   Gordon, "Primate Scream."

of "a latter-day Fagin" or "a James Bond who dispatches cadres on missions." In addition, "[f]ans see him as a cross between Lao Tzu and Che Guevara."[41] Alternatively, the baboons are comical human manques or mythologised for their apparently astonishing intelligence in infiltrating houses and fridges. (This surprise engendered by animal cognition is another repetitive response in media representation of animals.) When baboons embody agency and intentionality, as they do here, they are judged as disorderly and dangerous.

Fortunately, some informed concern about the fate of the beleaguered baboon troops in the Cape Peninsula does exist. Baboon monitors have been managed by Baboon Matters for over a decade, but sometimes at great risk to their safety. In August 2006, the *Cape Times* confirmed that two baboons had died of dieldrin poisoning.[42] Subsequently, Jenni Trethowan of Baboon Matters was poisoned after nursing the dying Angelina Ballerina in a rare example of a human sacrificing herself for an animal. Trethowan claimed that "the only positive spin-off of [her] poisoning incident was that it had drawn attention to the plight of the baboons."[43] The whimsical naming of the young female as well as the mention that she "died in Trethowan's arms"[44] contributes to the representation not only of the female baboon's individuality but also of her subjectivity and vulnerability. In taking responsibility for the deaths of these baboons, Jenni Trethowan heals, to some extent, what Derrida call the "abyssal rupture" between humans and other animals,[45] an aspect that is conveyed in the sympathetically written articles.

Another narrative featured a baboon who was called David, an aspect that some journalists seemed uncomfortable about, reporting that he had been "dubbed" or "nicknamed" David, as if the naming of an animal lacks legitimacy. David attempted to escape from his home troop of baboons in Tokai and to take control over his environment in order to find a mate in another troop. Given the human encroachment on his habitat, however, his quest was doomed. Trethowan regarded David as "an incredibly brave animal [who] tried three times to get across the [densely settled] Cape Flats, wanting to migrate to the Hottentots Holland [mountains]," a distance of about 60 kilometres.[46] Described as a "fugitive" and "on the run" like a criminal for daring to enter business and suburban areas,[47] he was photographed prolifically on rooftops and playing with the wires of a TV aerial. *People's Post*, a community newspaper, labelled the latter photograph "Bad Reception."[48] The *Tatler*, another community newspaper, went further, quipping in a caption to the photograph of David on a roof that he had driven the SPCA and

---

41   Ibid.
42   Gosling, "Banned Poison."
43   Gosling, "Poisoned 'Baboon Woman' Relapses."
44   "Body of Baboon Incinerated."
45   Derrida, "Animal Therefore I Am," 398.
46   Quoted in Gordon, "Primate Scream."
47   Collison and Prince, "Fugitive Baboon Evades."
48   Verbaan, "David the Baboon Darted."

police "bananas"; that he had been up to his "monkey business" and that, like a criminal, he was "still at large."[49] This tendency to represent animals comically, which constitutes one of the dominant discourses deployed in relation to animals in South Africa's newspapers, echoes some of the punning headlines in the Yengeni bull saga and the theatrical chicken killing, pointing to an inability to take animals seriously, with animals generally trivialised as cute or pitiful creatures.

David, finally, was darted but then had to be operated on for a gash on his leg. After recuperating at a Centre for Rehabilitation of Wildlife in Barrydale, he was relocated to the Limietberg Nature Reserve in the Dutoitskloof Mountains, but the narrative has a tragic closure. In what the SPCA Director Allan Perrins describes as an "untimely and horrible death," David was attacked by four dogs on a farm in a "fight to the death." The farm foreman had cut down the tree in which David had taken refuge after "apparently scavenging for food at two farm houses eighteen kilometers from where he was released."[50] The dryness of some of the SPCA comments about an "errant baboon," and Cape Nature Biodiversity Director Kas Hamman's comment that no one would be prosecuted, underscore the opinion that the baboon was out of place and imply that there is no solution for dispersing baboons. John Yeld's article on the death of David at least refers to him as "Cape Town's 'gentleman baboon.'"[51]

Perhaps one of the most distressing and uniquely South African examples of the naming of an animal in connection with a death industry is that of Frida, the lion cub who now exists with little hope of liberation from a game farm which has connections with the canned hunting fraternity. Canned hunting involves the breeding of predators, like lions, who are then "hunted" in enclosed spaces. As the writer of *The Sunday Independent* editorial bitterly put it: "In case this proves too difficult for some of the trophy hunters, the animals are sedated lest they spoil the fun by moving around before they can be shot."[52] The state, however, seems finally set to put an effective stop to canned hunting, but between 3,500 and 4,000 lions in the industry face a perilous future with fears of an impending "massive animal welfare crisis as breeders threaten to ditch unwanted predators that no longer have any commercial value"; in addition, the South African Predator Breeders' Association might appeal against the ruling.[53]

It has taken the fate of a European-bred lioness, a story of an individualised and named juvenile lion, to turn a particular kind of attention onto the unspeakable suffering of animals in this industry. Frida's life has comprised an extraordinarily convoluted narrative: rescued from a zoo in Bucharest in 2005, she was later sent via what her new owners assumed to be a trustworthy European-based animal

---

49   "Chilling Out."
50   Breytenbach, "SPCA to Investigate Killing."
51   Yeld, "David Killed by Four Dogs."
52   "Environmental Thugs Face Can."
53   Cadman, "Future No Better for Lions." For more information consult: http://www.cannedlion.co.za and http://www.lionrescue.org.za.

organisation, Vier Pfoten, to the Camorhi Game Lodge in South Africa.[54] In spite of being warned that this reserve was potentially a canned hunting lodge, the organisation went ahead with Frida's re-location there, and subsequently bought it from the owners. Vier Pfoten now refuses to release her to the Drakenstein Lion Park, which is a reputable sanctuary for rescued lions. Frida, as a young lioness, mauled the arm of a visitor to the farm, an entirely uncharacteristic act for a lion who had been habituated to humans.[55] The article effectively spells out the ramifications of each step of Frida's mismanagement by Vier Pfoten, making its point emotively with a photograph of a neotenous Frida facing the camera but with her ears turned back, which gives her a slightly anxious demeanour.

Another shorter article in a national weekly newspaper includes a triptych of photographs: Frida as a newly born cub behind bars in the dust of Bucharest Zoo, a close-up of the young cub, and then an older Frida with a Jack Russell terrier at Camorhi, now renamed Lionsrock.[56] The last-named photograph, however, does not hint at the horror of her life-sentence within a canned hunting reserve: she will be used as a breeding machine until she, too, will be killed by "sportsmen" in a country which "has achieved international notoriety as a premier destination for canned-lion hunters."[57] The articles on Frida in their focus on her vulnerability and youth and in their sense of the possibility of hope, small though it may be, may be more effective in motivating readers to take cognisance of the plight of animals within the canned hunting industry, than the more generalised reports which can only leave any sensitive readers with feelings of impotence and horror. Helen Bamford's excellent article which includes a shocking photograph of a hanging lion about to be skinned is a case in point.[58]

If Frida is photographed as a winsome and vulnerable cub, the discussion itself does not sentimentalise her in the othering way that the Yengeni bull was constructed. In Frida's case, sentimentality could even be regarded as a positive response if it arouses concern and raises awareness of the canned hunting industry. Philip Armstrong argues elsewhere in this volume that sentimentality can function as a powerful stimulus for change in relation to nonhuman animals, for "sentimental effects may mask the point at which a radical transformation of cultural feeling has taken place—or is about to take place." Certainly Armstrong convinces in relation to the narratives of named cetaceans Opo and Skana. Like the story of Frida, and unlike that of Yengeni's bull or Veronica's avian stunt double, they do not mask an underlying cultural hypocrisy towards a particular species of nonhuman animal.

Ironically, when the media attempts to represent the unspeakable, the effects may be contradictory, depending on specific instances. Thus the narrative of the individualised and named Frida, who becomes iconic of all lions within canned

---

[54]　　Van Riel, "Cub at Mercy."

[55]　　Ibid.

[56]　　Macleod, "Feud Over Frida."

[57]　　Cadman, "New Law."

[58]　　Bamford, "Canned Hunting."

hunting, foregrounds the ethics of "real" animals being located as sacrificial beings within modernity more powerfully and appealingly than discussion of the industry itself. Similarly, the story of the baboons' poisoning attests to their unspeakable treatment in peri-urban areas but, because their suffering is individualised, Angelina Ballerina is named, and their human champion was seriously ill, the narratives effectively show the very specific outcomes of the lack of human compassion, surely because the reading public, generally, is not directly implicated in their deaths. In the instances of the killing of domestic animals, however—the bull sacrificed by Yengeni and the chicken killed in *iMumbo Jumbo*, both unnamed but individualised—the unspeakable practices within which they are inserted have to be denied, because meat-eating is culturally accepted. What is too challenging for readers, and for most writers in the media, I would suggest, is the imperative to respond to the (mostly) unasked Derridean question about whether we have a responsibility to the living. The sacrificial discourses of modernity and tradition remain only too mundanely acceptable as the noncriminalised killing of animals continues.

Chapter 5
Possum Magic, Possum Menace: Wildlife Control and the Demonisation of Cuteness[1]

Kay Milton

The idea for this essay came from a casual observation of an obvious fact: that Australians and New Zealanders hold very different attitudes to possums, more specifically to the common brushtail possum (*Trichosurus vulpecula*),[2] a native of Australia which has become widespread in New Zealand. Australians have a love-hate relationship with them. They are seen as appealing and entertaining, and are valued as a component of the country's native fauna, but they can be an intolerable nuisance when they invade homes and gardens. New Zealanders appear to loathe them as one of the worst threats to their ecology. On the face of it, the reasons for this difference are straightforward. Like many other species, possums are valued as long as they stay where they belong, in the Australian bush. When they become "matter out of place"[3] by invading spaces where they do not belong, and especially when they do so destructively, they can be seen as a menace. But a dislike of possums may not be so simple. There is a widespread understanding among some scholars that humans are "naturally" inclined to like cute furry animals. If this is so, specific cultural mechanisms may be needed to make people dislike possums when it is deemed necessary that they do so. This essay compares the representation of possums in Australia and New Zealand in order to examine how New Zealand culture demonises possums, thereby overriding or ignoring any appeal their cuteness might have. It draws on material from the news media, government and NGO publications, websites, and conversations with conservationists and others in Australia and New Zealand.

---

[1]   I am grateful for comments on earlier drafts of this essay from Annie Potts, Jane Mulcock, members of the Department of Anthropology, University of Auckland, and members of the Department of Anthropology, University of Queensland.

[2]   Hereafter, when the term "possum" is used, it refers to the common brushtail possum unless otherwise stated.

[3]   Douglas, *Purity and Danger*.

The "Cute Response"

The "cute response," which is considered by psychologists to be more or less universal in human beings, is the feeling people experience on encountering beings whose appearance stimulates a desire to hold, cuddle, and protect them. It belongs to a category of mechanisms known as "social releasers," because they "release" or induce social behaviour, in this case parenting behaviour. The strongest stimuli for the cute response are therefore expected to be human babies. John Archer has pointed out that,

> The possible existence of social releasers in humans has been hotly debated. There is, however, considerable agreement (and considerable evidence ...) that humans respond in a parental way to certain sets of facial and bodily features found in human infants. These features make most of us go "ah" and "coo," and regard their owner as "cute" or "sweet."[4]

In an earlier publication Archer provided the evidence he refers to here by summarising the results of studies showing that human adults, and especially women, respond positively to baby-like features.[5] Konrad Lorenz, who studied parenting behaviour in a number of species, observed that the young of different animals often have similar physical features—round faces, relatively large eyes, short noses, short thick limbs, and clumsy movements.[6] Some adult animals retain these features, making them more likely to provoke a cute response in humans.

   This is not the place to discuss whether the cute response is a "natural" human reaction. In general, I do not consider it useful to classify human responses as "natural" or "cultural." There appears to be broad agreement that the appeal of cuteness, whatever its basis, is widespread and varies in intensity across individuals and cultures. In the pet-keeping cultures of Europe and America, the appeal of cuteness has literally shaped some animals[7] which have been deliberately bred to exaggerate baby-like features. The stereotypical "pampered pets," those expected to be "mothered" by their owners, are small fluffy dogs (Pekinese, toy poodle, Pomeranian), Persian cats, and other breeds of similarly cuddly appearance.
In Japan, where the cuteness of children is heavily emphasised, visitors to monkey parks are particularly attracted to young monkeys by their cute appearance and responsive nature,[8] and koala cuddling is said to be especially popular among Japanese visitors to Australia.[9] In Bengkulu, Indonesia, where biological anthropologist Dan Fessler carried out research, there is a word, *gelinggaman*, which has no direct equivalent in English, and which describes the feeling of being

---

4   Archer, "Why do People," 249.
5   Archer, *Ethology and Human Development*, 82–90.
6   Archer, "Why do People," 249.
7   Serpell, "Anthropomorphism," 83–100.
8   Knight, "Maternal Feelings," 186–7.
9   Knight, pers. comm, February 2008.

"powerfully attracted by cuteness."[10] Informal conversations with colleagues have revealed other such instances in South-East Asian languages.

It is remarkable how frequently animals are judged in terms of the cuteness of their appearance, even in contexts in which it might be thought irrelevant. The following quotation comes from an article on climate change in *The Weekend Australian*: "In Australia, highly adaptable species such as cane toads, cockroaches and some kangaroo species would probably cope, said Andy Pitman, a climate scientist with the University of NSW. 'But species that lack tolerance to climate like some small possums and koalas—*the cute ones*—would not (survive)'" (emphasis added).[11] Here cuteness is mentioned in a conversation about biodiversity, concern for which is often explicitly indifferent to people's emotional responses to animals. The implicit reasoning is that people like cute animals and consider them worth conserving, so they are more likely to be concerned about global warming if it threatens cute species than if it threatens unappealing ones like cockroaches and cane toads.

If cuteness is considered an important feature of animals in general, what of the brushtail possum? Does it have cute appeal? It possesses some of the appropriate physical features: it has large, forward-facing eyes, and a roundish face when viewed from the front; it is soft and furry, about the size of a cat, has short thick limbs and, on a flat surface, moves clumsily: "On the flat they look a bit like a toddler in a nappy."[12] Its cuddliness is enhanced by its shape; its front limbs resemble arms and its paws are like hands, and a young possum holds onto its mother much as a human child does. In a study of public attitudes to possums in Sydney, 68.8 per cent of respondents (n = 125) agreed that possums are cute.[13] In a public information leaflet, the Department of Conservation and Land Management (CALM) in Western Australia describes possums in general as "delightful and appealing creatures, with their soft downy fur and large innocent eyes."[14] But attitudes to possums, and to any animal, are never shaped purely by their appearance. People's experience of possums, mediated by cultural meaning, determines whether they are loved or despised. The following sections describe how possums are experienced and thought about in Australia and New Zealand.

Possums in Australia

The common brushtail is one of twenty-six species of possum native to Australia. It is highly adaptable and was once widespread, but like many Australian

---

[10]   The literal meaning of this word is given as "shivers up the spine," and it is also translated as "the willies [i.e., experience of the uncanny]." Fessler, "Shame in Two Cultures," 215.

[11]   Dayton, "Climate."

[12]   Nattrass, *Talking Wildlife*, 10.

[13]   Hill, Carbery, and Deane, "Human-Possum Conflict," 106.

[14]   Department of Conservation and Land Management, *Encouraging Possums*, 1.

mammals, it has suffered predation by introduced carnivores (foxes, cats, dogs) and destruction of its natural habitats. Its numbers were severely reduced by hunting for fur during the latter decades of the nineteenth century, with several million being killed annually.[15] As a result it has disappeared from more than half of its former range and is common only on large islands and in some cities.[16] Its adaptability has enabled it to move into urban and suburban areas where it is frequently encountered by people. Exotic plants in parks and gardens are especially attractive to possums because, unlike native Australian vegetation, they do not produce the toxins that deter grazing and so limit damage. The roof-spaces of suburban homes are attractive to possums as dens for sleeping during daylight hours. They can be extremely noisy when leaving or entering these dens, when walking across roofs and especially when fighting with other possums. Along with their liking for garden plants, a tendency for dogs to bark persistently at them, and the mess and smell caused by their droppings, this brings them into conflict with human beings.[17]

Surveys of public attitudes to possums in urban and suburban areas have revealed mixed feelings about them. In a questionnaire survey in the City of Knox, Victoria, in the late 1990s, 25.4 per cent of respondents (n = 142) expressed a positive attitude to possums and 33.1 per cent expressed a negative attitude (the remaining respondents were neutral or did not answer the attitude questions).[18] In a more recent survey, conducted in the Mosman area of Sydney,[19] 54.8 per cent of respondents said that possums were welcome on their property, 58.9 per cent thought they were an important part of the urban environment, 46.7 per cent thought they were (or would be) nice to have around, and 70.5 per cent thought they should be allowed to live in urban areas (n = 122–126). Between 30 and 40 per cent of respondents thought possums were a nuisance, made too much noise, and caused too much damage to trees and garden plants, while 49.2 per cent thought they made too much mess with excrement; accordingly, 87.8 per cent felt that homeowners should be allowed to remove problem possums from their properties (n = 119–126).[20]

In view of the legal obligation to protect possums, conservation authorities try to educate the public on how to co-exist peacefully with them. The Department of Natural Resources and Environment (DNRE) in Victoria issued a 12-page booklet in the 1990s, which gives instructions on how to trap possums for removal from roof-spaces—the only circumstances in which this can be done legally without a licence.[21] Once trapped, the possum must be released at sunset on the same day

---

[15]   Matthews et al., "Brushtail Possums," 159.
[16]   Ibid., 160.
[17]   Temby, *Wild Neighbours*, 173.
[18]   Miller, Brown, and Temby, "Attitudes Towards Possums," 121.
[19]   Hill, Carbery, and Deane, "Human-Possum Conflict."
[20]   Ibid., 106.
[21]   Department of Natural Resources and Environment, *Living with Possums*.

and within 50 metres of the trapping site (the roof having been made possum-proof in the meantime). Research has shown that possums have a poor rate of survival if removed from their home ranges.[22] The public are encouraged to provide boxes attached to trees as alternative dens. CALM in Western Australia issues information notes on "Encouraging Possums" under their Land for Wildlife Scheme. The tone is more positive than in the Victorian document: "If you are fortunate enough to have possums [of any species] living in your area or on your property, you need only to make sure that the habitat remains as suitable as possible for them to stay."[23] This positive approach may reflect the relative scarcity of possums in Western Australia vis-à-vis the eastern states. The CALM leaflet acknowledges that it may be necessary to remove possums from roofs, but any wish to remove them from the area is discouraged with the words, "… hours of potential pleasure watching their antics outside the house will be lost!"[24]

Feelings about possums sometimes run high enough to make them a political issue. In Melbourne in 2004, municipal authorities became concerned that possums might spread a deadly wilting disease among the city's palm trees. They installed metal tree protectors to prevent the possums climbing the trees. This angered protesters, who feared that the metal bands might trap possums already in the trees, causing them to die of starvation or in their attempts to jump over the bands. This issue flared up again in St Kilda early in 2007, where both the palm trees and the possums are visitor attractions. The City of Port Phillip installed tree guards, which were taken down by protesters. The City authorities planned to reinstall them, but said they would check for possums first and only attach guards to trees which had no possums in them.[25] According to Protectors of Public Lands Victoria (a coalition of community action groups and environmental organisations), Port Phillip Council was trapping and killing the St Kilda possums on the mistaken assumption that they were killing the palms by chewing their fronds. An invitation to a public meeting in February 2007 urged people to "Come to the Catani Gardens in St Kilda to meet, greet and feed the possums—help rescue them from Death Row …."[26]

It would appear from this brief summary that Australian attitudes to possums are mixed. They are respected and protected as members of the nation's native fauna, and although their adaptation to urban and suburban areas brings them into conflict with some, many people are pleased to have them around.

---

22 Matthews et al., "Brushtail Possums," 164–5.

23 Department of Conservation and Land Management, *Encouraging Possums*, 3.

24 Ibid., 4.

25 Australian Broadcasting Commission. http://www.abc.net.au/melbourne/stories/s1569606.htm (accessed 1 May 2007; page now discontinued).

26 Protectors of Public Lands Victoria. http://www.protectorsofpubliclandsvic.com/ (accessed 1 May 2007; page now discontinued).

Possums in New Zealand

Brushtail possums were introduced into New Zealand in the nineteenth century to establish a fur trade.[27] Repeated introductions by acclimatisation societies assisted their spread across the country. Hunting of possums was at first unregulated but concerns about possible "over-harvesting," expressed by acclimatisation societies, led to a ban on unlicenced hunting in the early twentieth century. By then, farmers and horticulturalists were becoming concerned that possums were damaging crops. A report published in 1920 recommended that possums be released in forests, away from orchards and gardens. During the next three decades, concern grew about the impact of possums on native forests. This resulted in a radical change in government policy from 1947: all restrictions on taking possums were cancelled, penalties were introduced for releasing them, and the use of poison to control their numbers was legalised. From 1951 to 1961 bounties were offered for possums skins to encourage their control, but this failed to have a significant impact and may even have stimulated illegal releases. From the 1960s onwards, legislation has been introduced to control possum numbers on agricultural land (Agricultural Pests Destruction Act 1967), on crown land (Wild Animal Control Act 1977), and locally under Pest Management Strategies (Biosecurity Act 1993). Despite this, possums have continued to spread to all habitable areas of New Zealand, and their numbers are estimated at around seventy million.[28]

While Australian websites and public information leaflets educate people on how to live with possums, the equivalent media in New Zealand describe and recommend efforts to get rid of them. Visitors to forest reserves are warned about poison baits set for possums. The Department of Conservation website gives instructions to the general public on how to deprive possums of nest sites, how to protect trees using metal strips and home-made repellents, and how to kill possums using poison bait and traps.[29]

Ironically, the fur trade, which brought possums into New Zealand in the first place, is now assisting in their control by providing a market for possum products. Possum fur is marketed as "eco-friendly fur," the sale of which is supported by environmental organisations. The conservation director of the Worldwide Fund for Nature (WWF) New Zealand was quoted as saying, "We support killing possums. We'd like to see all possums as dead possums, and the fur industry may provide an additional incentive for people to kill possums."[30] The public, including tourists to New Zealand, are left in no doubt that they are doing the country a big favour if they buy possum products.

---

[27] General information on possums in New Zealand is summarised from Clout and Erikson, "Anatomy of a Disastrous Success."

[28] Kiwi Conservation Club (KCC), "Possums."

[29] Department of Conservation, "Possums."

[30] Quoted in Klotz, "Fur Fashion."

Cuteness and Conservation

What is striking about the New Zealand discourse on possums, as an expression of human-animal relations, is the relative lack of concern for them as individuals. This is not totally absent; some methods of killing are favoured "from a welfare perspective"[31] because they cause the least suffering. But it is never allowed to override the necessity of killing them for conservation reasons. Does the relative absence of sympathy for possums in New Zealand mean that New Zealanders, unlike Australians, are indifferent to their cuteness? A maker of possum pies was quoted as saying, "They might be cute and fluffy over in Australia but they are certainly not cute and fluffy here."[32] But the Department of Conservation acknowledges their cuteness: "The possum is high on cuteness, and equally high in nuisance value."[33] In a research project on public perceptions of possums in New Zealand, participants in focus groups thought possums were "a major pest," but also saw them as, "an aesthetically pleasing animal (using language such as 'soft,' 'cuddly,' 'fluffy,' 'cute' and 'nice brown eyes')."[34] This suggests that at least some New Zealanders are responsive to the cute appearance of possums, but any sympathy that may be provoked by this is overridden by other considerations. This highlights a difference between New Zealand and other countries in terms of how public support for conservation can be encouraged.

In many parts of the world, cute animals are a valuable resource for conservationists. NGOs are well known for using images of appealing and attractive animals to encourage public support for their campaigns. The WWF's panda logo is the best known example, but there are many others. In the logo of the International Fund for Animal Welfare (IFAW), protective human hands encircle a seal, an echo of the campaign in the 1970s to ban the hunting of young harp seals in Canada. Thousands of Europeans were induced to support a ban on the import of seal products into Europe by photos of white fluffy seal pups with large black appealing eyes.

Conservationists in New Zealand do not have the option of using cuddly, wide-eyed mammals in their campaigns. All mammals, except seals and bats, are alien to New Zealand, and many of them, like the possum, have had a damaging effect on indigenous wildlife. On a public information poster in Puketi Forest (Northland), an animal shown being cradled protectively in human hands is not a wide-eyed seal, but a kiwi; with tiny eyes and a long, tube-like bill, it is almost the antithesis of Lorenz's classic images of cuteness. So, in the context of theories about emotional responses to animals, New Zealand has a problem in its efforts to conserve its natural heritage. The invasive introduced species that are destroying

---

[31] Eason, Warburton, and Henderson, "Toxicants Used for Possum Control," 162.

[32] Stuff.co.nz. http://www.stuff.co.nz/stuff/0,2106,2954125a11,00.html (accessed 20 June 2004; page discontinued).

[33] Department of Conservation, "Facts about Possums."

[34] Fitzgerald, Wilkinson, and Saunders, "Public Perceptions," 162.

local ecosystems and endangering indigenous wildlife tend to be more appealing in their appearance than the native species are, and the possum is one of the cutest. The anti-possum discourse in New Zealand overrides any cuteness appeal they might have by casting them in two negatively valued roles: as disease carriers and as undesirable aliens.

Possums as Carriers of Disease

Disgust is considered by many psychologists to be a universal and basic human emotion[35] which has evolved as a protection mechanism against disease. There is an expectation among evolutionary psychologists that humans will be repelled by substances and objects likely to cause disease, such as blood, semen, saliva, faeces, and urine. Following the work of Douglas[36] and others, it is well understood by social and cultural anthropologists that such substances attract a lot of symbolic attention. They are controlled by rules of hygiene and avoidance, there are "proper" places and methods for dealing with them, which reduce the danger of contamination. Leaving aside the contentious issue of whether such responses should be seen as "natural" or "cultural," their more or less universal occurrence leads to the expectation that the association of something with disease will be a powerful stimulant for negative emotions. It immediately becomes potentially contaminating, something that should be avoided or eliminated.

Since the 1960s, possums in New Zealand have been identified as carriers of bovine Tuberculosis (Tb). The disease was introduced in cattle brought by early European settlers. It is similar to human Tb, and can be caught by humans, for whom it can be fatal; this risk has been largely eliminated since the pasteurisation of milk was introduced. Nevertheless, bovine Tb remains a serious threat to livestock and to the economic health of cattle and deer farming in New Zealand. A compulsory Tb control programme now helps to minimise the threat, and part of this programme involves controlling the numbers of wild animals known to carry bovine Tb. In the 1960s, chronic Tb in cattle was attributed to a high rate of infection in the local possum population. Infected possums have since spread to about 38 per cent of the country and "Scientists now regard possums as the most important vector of Tb, for most infected herds in New Zealand, over the past 25 years."[37]

I would not suggest that fear of contamination is the primary reason why New Zealanders apparently have no difficulty in disliking possums—attempts to eradicate possums were in place before it was discovered that they carry Tb—but it must make it easier to dislike them. It places them, effectively, in the same category as rats, which must be one of the most widely despised animals in Western cultures, historically famous as vectors of bubonic plague.

---

[35] Evans, *Emotion*.
[36] Douglas, *Purity and Danger*.
[37] Green, *Use of 1080*, 16.

Possums as Undesirable Aliens

If the fear of contamination from possums is a fairly basic emotional response, their rejection as unwanted aliens in New Zealand depends on a complex assemblage of cultural conditions. These include the status of New Zealand as an independent and socially responsible country, and a range of ideas and assumptions that inform the practice and discourse of nature conservation worldwide. Every document I have read on possums in New Zealand mentions that they are not native to the country but were deliberately introduced by people. Many do so in the opening lines, creating the impression that it is considered one of the most important things to say about possums, as if their foreignness outweighs their other characteristics and determines their value.[38]

It is well documented in the historical and social science literature that awareness of and identification with the natural environment is often a part of the process through which emerging nations gain their identity and independence. For instance, the designation of "national" parks, as areas within which a country's "natural heritage" is protected, has been seen as an assertion of national identity, particularly in colonial and postcolonial contexts.[39] Although the purpose of such designations has often been to protect scenic landscapes, the conservation of ecosystems has also been a major objective. In 1992, the nations represented at the Earth Summit in Rio de Janeiro adopted the Convention on Biological Diversity, which formally recognised the responsibility of each nation state to protect the biodiversity within its borders. The extent to which a nation is able and willing to do this contributes significantly to its international reputation. This strengthens the perceived need to reduce the threat to ecosystems from introduced species. The brushtail possum, in damaging New Zealand's native forests, is not only destroying something that belongs uniquely to New Zealand and helps to shape the nation's identity, but also calls into question New Zealand's international reputation as a responsible country. Regardless of whether it is possible to eradicate the brushtail possum, the New Zealand authorities must be seen to be doing what they can to reduce its numbers.

At the root of the international discourse on the conservation of biodiversity is a sharp distinction between human and nonhuman processes. Biodiversity is treated as a natural phenomenon, one that arose independently of human activity, even if it now depends on humans for its conservation. In this context, it is important that possums and other unwanted species were introduced into New Zealand by people. Had they got there without human assistance, they might

---

[38] In addition to the official, science-driven discourse that encourages New Zealanders to think of possums as undesirable aliens, there is a parallel popular discourse that belongs to the (mostly) good-natured rivalry between New Zealand and Australia. I was told several times during conversations about possums in New Zealand, though never in total seriousness, that possums are hated because they are Australian.

[39] For examples see Ranger, *Voices From the Rocks*; and Carruthers, *Kruger National Park*.

still be considered a menace, but their control would have to be justified in other terms, and it is questionable whether complete eradication would be considered an ideal. This can be highlighted through comparisons with other countries in which native species are managed. In Australia, as we have seen, possums are often a serious nuisance, but attempts to control their numbers can provoke public protest. Similarly, in the UK, the badger, like the possum, is a carrier of bovine Tb, but the culling of badgers has always been highly controversial, partly because they are native animals. Groups dedicated to the protection of British wildlife, such as the many regional Wildlife Trusts, are heavily involved in the debate about badgers and bovine Tb, questioning whether badgers infect cattle or vice versa, and arguing that culling may not be the most effective method of controlling the spread of the disease. In North America, beavers, like possums in New Zealand, can seriously damage native trees. But, again because they are a native species, their control is controversial.

Possums and the Socialisation of Children

The negative images of possums described above are supported largely by the policies and practices of conservation and agriculture, which are predominantly (though not exclusively) adult domains.[40] This raises the question of how children are socialised into thinking of possums in negative terms. In a comparative analysis of folk biology, Coley identifies three important questions: "First, what are children's conceptions of living things like? Second, what are adults' conceptions of living things like? And third, how do we best characterize the process by which children's conceptions come to resemble those of adults in their culture?"[41]

This line of questioning is pertinent to the cultural representation of possums in New Zealand and Australia. If Australian adults hold broadly favourable (though often mixed) attitudes to possums, and New Zealanders are wholeheartedly against them, one can expect something in the socialisation of children to generate these cultural differences. And if the appeal of cuteness is as widespread as some scholars suggest, one might expect to find mechanisms that explicitly encourage or discourage it. I have not tested this expectation through systematic research, but even casual observations point to striking differences between the two countries.

For instance, in major bookstores in Australian cities, the children's section includes many books featuring mammals, including possums, as the main characters. *Possum Magic*[42] is one of the best loved Australian children's books of all time, undergoing 32 reprints in its first 20 years. A "21st birthday edition" published in

---

[40]   Conservation organisations in New Zealand, as elsewhere, play an educational role. For instance, the Kiwi Conservation Club (KCC) website explains the possum problem in a direct but light-hearted fashion, using alliterative language (possums are "pesky pests"), cartoon drawings, and vivid comparisons.

[41]   Coley, "On the Importance of Comparative Research," 85.

[42]   Fox and Vivas, *Possum Magic*.

2004 was reprinted six times in its first three years, and the book was the subject of a nation-wide online "book rap" for primary schools in 2004. In contrast, the main bookstores in Auckland display very few children's books featuring wild mammals. Books about kiwis are popular,[43] as are books about domestic animals.[44] But possums feature very rarely in children's stories, and when they do they are forces of destruction. For instance, in *As Kuku Slept* possums are shown with angry expressions on their faces, eating their way through the forest and threatening the lives of the benign, endemic kereru (New Zealand woodpigeon).[45] Their role is as unambiguous as that of the "big bad wolf" in traditional European fairytales.

Children's literature forms just a small part of the socialisation process. To address Coley's questions fully we would need to examine the messages children receive from their families, from their teachers, through the media, and from each other. And to discover whether these messages build upon or suppress a prior inclination to be attracted by cuteness we would need a great deal more research.

Concluding Comments

When I presented an early draft of this essay at the University of Queensland, a student who had been brought up in New Zealand commented, "You know, it's very strange. Possums in Australia are cute, but possums in New Zealand are not cute, and yet it's the same animal!" Asking how people come to perceive "the same animal" in completely different ways is an important step towards understanding how cultural conceptions of animals in general are created and sustained. Clearly, the animals themselves play a major role in shaping those cultural conceptions, albeit (one assumes) unwittingly. How they look (cute or ugly), what they do, where they do it, how their activities come into contact with human ones, are fundamental to cultural representations of animals. But it is important also to understand the selective processes through which cultures shape animals, criminalising or sanctifying them, making them objects of disgust and contempt or concern and protection, in accordance with changing priorities. Cultural priorities determine whether animals are in the wrong place (possums in New Zealand, possums in Australian roof spaces) or the right place (possums in the Australian bush) and shape their treatment by human society.

---

[43] Owen and Gunson, *How the Kiwi Lost Its Wings*; Darroch, *The Kiwi Who Lost His Mum*.

[44] For instance, Lynley Dodd's *Hairy Maclary from Donaldson's Diary* and *Slinky Malinki*, which were the first books in two highly successful series.

[45] Devlin, *As Kuku Slept*.

PART 2
Ethics

Chapter 6
Pleasure's Moral Worth

Jonathan Balcombe

Introduction

The academic study of animal pleasure has no name, for it is, as yet, a mostly neglected field. Today, there are at least 20 scholarly journals dedicated to the study of pain, yet none that focus on the study of pleasurable experience. This is curious, for while the cessation of pain and suffering may be more urgent for the individual experiencing them, the pursuit of pleasure is not only important to individual health and well-being, but also drives much of what humans—and as I shall argue, nonhumans—do in their daily lives.

As an ethologist interested in the study and implications of how animals experience their worlds, I am hopeful that pleasure—in both humans and nonhumans—will gain prominence as a legitimate subject of scientific inquiry. I have proposed elsewhere that this field be called Hedonic Ethology.[1] The adjective "hedonic" derives from the Greek *hedone*, meaning pleasure.

The main goal of this chapter is to argue for pleasure's moral significance. I will first provide some historical background which helps explain why pleasure has—with a few notable exceptions—been broadly excluded from scientific inquiry. Next, I will present some basic arguments as to why the notion of animal pleasure should be no more controversial than that of human pleasure. I will follow this with a few examples of animal pleasure. I close with a discussion of the moral significance of pleasure—in particular, how it informs the human-animal relationship.

Some readers may feel that a definition of "pleasure" is warranted. However, such is the familiarity of this concept that most everyone has a good grasp of its meaning. Let me just say that I intend it as a generic, umbrella term for any sort of positive experience whose basis could be physical, emotional, or both.

Historical Background

Science's general neglect of the study of animal pleasure falls within a broader reluctance to study aspects of animal experience that may be termed "private." Because nonhumans cannot express their feelings verbally,[2] our interpretation of

---

[1]   Balcombe, *Pleasurable Kingdom*.

[2]   Studies with great apes taught to use sign language or other forms of symbolic communication have partially eroded this language barrier.

their feelings is hampered somewhat. Science has traditionally adhered to Occam's razor, a principle which holds that in explaining a thing no more assumptions should be made than are necessary. Because we cannot actually feel what an animal is feeling, the assertion that the animal feels something depends on making an assumption that s/he is capable of feelings.

During the late nineteenth century, Occam's razor was not being applied as stringently as it would be in the twentieth century. Charles Darwin's 1872 book *The Expression of the Emotions in Man and Animals* used a largely comparative approach, presenting observations and anecdotes to argue for the emotional experiences of nonhumans based on parallels between human and nonhuman subjects. Early in the twentieth century, stimulated in part by Ivan Pavlov's celebrated studies of conditioned reflexes in dogs, behaviourism emerged as the prevailing scientific paradigm. Behaviourism holds that behaviour can and should be studied without appeal to thoughts or feelings. Under the behaviourist mantle, animal cognition and emotion were deemed taboo. It was not until the 1970s, when the American ethologist Donald Griffin wrote the first of a series of books on animal thinking, that the study of animals' mental and emotional experiences became once again acceptable.[3]

There is an inherent double standard in denying animals feelings because they cannot be incontrovertibly proven to exist. The same can be said of human feelings, which are no less private in the absolute sense. The solipsist claim—that the self is all that exists or can be known because we can only verify our own experiences and no one else's—is logically irrefutable. Yet, we accept without question the feelings of other humans. As we should. We have good reasons for refusing to apply Occam's razor to the assumption that humans are sentient. The evidence is overwhelming; arguably, denial would make many more assumptions against what the evidence shows us.

Increasingly, scientists take the same position regarding animal feelings. Today, few scientists are willing to deny that vertebrate animals are sentient, though there are legitimate discussions of where to draw the line. The fortunate consequence of this more open-minded thinking is that science is now applying creative methods to the study of animal cognition, awareness, and emotion. In the case of pleasure, how an animal behaves, and the context of the behaviour, are useful clues as to how the individual may be feeling. Allowing an animal to make choices among a series of putatively desirable options, for example, is a useful tool for assessing what animals like. We can also measure physiological changes, such as the release of endorphins or changes in heart rate or blood pressure. In all of these applications, we may use the human response as a reference point, bearing in mind that our own preferences and responses may not always be a reliable predictor of those of another species.

---

[3]   Griffin, *Question of Animal Awareness* and *Animal Minds.*

Arguments for Animal Pleasure

Before reviewing a few examples as evidence for animals' experience of pleasure, it is necessary to briefly present four bases for why we should expect that the realms of pleasure are not confined to humans. Some readers may wonder why it should be necessary to build the case that certain nonhumans feel something as basic as pleasure, but as noted earlier, this topic remains nascent and largely neglected in scientific discourse. Thus, until animal pleasure is broadly and academically accepted, a detailed defence of its presence is warranted.

Pleasure is Adaptive

Perhaps the most fundamental basis for accepting animal pleasure is that it is useful in the evolutionary sense. Some billion years ago, when organisms first became mobile, an adaptive premium was suddenly placed on the possession of sensory systems. Being able to perceive and respond actively to one's environment allowed one to orient towards favourable things such as food sources and others of one's kind (especially useful for sexual reproduction), and to avoid aversive things such as solid objects that could cause injury or things that might eat you. Thus, the evolution of motility was a key step towards the eventual evolution of complex sensory systems that could differentiate noxious from rewarding stimuli. Just as the capacity for pain is adaptive for a mobile organism that can move away from aversive things, so too is pleasure beneficial by rewarding the individual for performing behaviours that promote survival and procreation.[4]

Animals Are Equipped to Feel Pleasure

For pleasure to be adaptive, animals need the physical equipment to experience it. All members of the vertebrate animal kingdom—mammals, birds, reptiles, amphibians, and fishes—share the same basic body plan: skeletal, muscular, nervous, circulatory, digestive, excretory, endocrine, and reproductive systems. To this shared foundation we can include a sensory system. All the vertebrates have the five basic human senses: sight, smell, hearing, touch, and taste. The senses function as the interface between an animal's nervous system and its surroundings. Equipped with the ability to detect and avoid unpleasant stimuli and to seek rewards, animals have the raw materials on which natural selection can act to favour pleasure and pain.

Humans and animals also share much of the same physiological and biochemical responses to sensory events. When we experience painful or pleasurable sensations, our brains send signals and our glands secrete compounds that help us deal with the situation. Human emotions are linked to two brain structures, the amygdala and the hypothalamus. Our positive feelings are also mediated by such biochemicals as

---

4   Cabanac, "Physiological Role of Pleasure," 1103–7.

dopamine, seratonin, and oxytocin. Many animals, especially mammals, possess these same neurological structures and brain chemicals. Imaging technologies such as PET and MRI provide further evidence that animals experience emotions like we do. There are, for example, remarkable similarities in active brain regions in guinea pigs experiencing separation distress and in human brains during feelings of sadness.[5] Animals' brains appear to be "wired" to respond to discrete types of sensory pleasure, including food pleasure, drug pleasure, and sex pleasure.[6]

Humans Feel Pleasure

The case for animal pleasure would be less robust were it not for the fact that humans experience pleasure. That we know and accept the existence of this sensory phenomenon in one species provides a firm foundation for its existence in others. That human languages contain rich vocabularies for describing good feelings attests to the diversity of both physical and emotional pleasures that can be felt by humans. English, for example, includes words such as: happiness, delight, surprise, anticipation, pride, satisfaction, joy, elation, ecstasy, thrill, euphoria, exultation, jubilation, excitement, rapture, fulfilment, gratification, and comfort.

We may not assume, either, that the range of pleasurable experiences felt by animals is delimited by those humans are capable of feeling. It could be that our complex social networks and sophisticated language have given rise to subtle emotions—such as satisfaction and gloating—that are either absent or poorly developed in other taxa. But to the degree that feelings derive from sensory systems, other animals may experience forms of pleasure inaccessible to humans. Having evolved in diverse environments, where different niches present novel adaptive challenges and opportunities, animals have varying sensory and perceptual skill sets. For example, bats and cetaceans orient by echolocation; some fishes communicate with pulses of electricity; birds and fishes can see a broader spectrum of colours thanks to their having four types of photopigments compared to our three; and sea lions can track and catch fish in unlit waters with their whiskers by detecting turbulence trails they leave.[7] Some animals can tune into the earth's magnetic field to help navigate, and there is now evidence that birds perceive this visually.[8] These examples do not explicitly involve pleasure, but they do illustrate the potential for pleasures unknown to us.

Animals Feel Pain and Distress

Pain in animals is well studied and well established. For animals with highly developed nervous and sensory systems, being able to learn from an earlier painful

---

[5] Panksepp, *Affective Neuroscience*.

[6] Berridge and Kringelbach, "Affective Neuroscience of Pleasure," 457–80.

[7] Dehnhardt et al., "Hydrodynamic Trail-Following in Harbor Seals," 29–31.

[8] Heyers et al., "A Visual Pathway," e937.

experience is a further refinement of the foundational stimulus-response dynamic. Cognitive creatures can recognise and remember the source of an earlier hurt, and make adjustments to their behaviour to reduce the chances of a repeat experience. The conscious pain perceiver is more likely to survive future encounters and to be favoured by natural selection.

The perception of noxious stimuli and their conduction to parts of the brain that register pain are fairly identical processes among the different mammals that have been examined. The benefit of measuring pain perception at the brain level is that it is not vulnerable to confounding factors such as stoicism, which may make an animal appear to be less in pain than she actually is. With regard to experiencing pain, there are no unequivocally "higher" or "lower" sentient species, at least among mammals.[9]

Indirect evidence for animal pain is also widespread in nature. Plants, for example, have exploited animals' capacity for pain and discomfort with the evolution of thorns and bitter-tasting chemical compounds in their tissues. Similarly, many animals have well-developed spines, stingers, horns, and tusks which, like thorns, not only inflict pain, but also signal "don't touch." More direct evidence for animal pain resides in animals' behavioural responses to noxious stimuli. They call out in pain, they avoid and retreat from sources of pain, and they flinch, limp, and protect the injured part. These responses are all compellingly consistent with the sentient experience of pain.

Scientific studies of animal pain reinforce these observations. Laboratory studies have shown repeatedly that injured rats, for example, will favour the bitter taste of water that contains a pain-relieving drug over unadulterated water.[10] Commercially raised chickens often experience leg problems associated with selective breeding, growth, and husbandry, and can develop highly abnormal gaits or become completely unable to walk. Lame birds presented with food laced with painkillers or unadulterated food ingest more of the drugged food than do non-lame birds, and as the severity of the lameness increases, lame birds consume a greater proportion of the drugged food.[11] Empirical evidence is now emerging that suggests that some invertebrates experience pain, which raises the intriguing possibility that they may feel pleasure also.[12] For instance, a study of captive crayfish found that they prefer to swim in a quadrant of a test arena into which was infused solutions of cocaine or amphetamine.[13]

Similarly to the study of pain, stress, and distress (which may be defined as stress that persists to the point of dysfunction) have been well studied in animals, especially laboratory-housed rodents. It has long been observed that social isolation is deleterious for rats, and that so-called "isolation stress" alters physiological and

---

[9]   Silverman, "Sentience and Sensation," 465–7.
[10] Persinger, "Rats' Preferences," 674–80.
[11] Danbury et al., "Self-Selection," 307–11.
[12] Barr et al., "Nociception," 745–51.
[13] Panksepp and Huber, "Ethological Analyses," 171–80.

86 *Considering Animals / Balcombe*

behavioural characteristics.[14] Rats housed alone were deemed more stressed than rats housed four per cage, as judged by significantly higher heart rates and blood pressures recorded in the solitary rats.[15] Behavioural symptoms of isolation stress in mice include aggression, stereotypies (prolonged repetitive, fruitless behaviours such as gnawing at cage bars, or performing somersaults), convulsions, nervousness, and handling difficulty. Physiological symptoms include lower immunity, higher tumour incidence, ulcers, and increased pathology.[16] Mice housed in barren cages drank significantly more water that contained an anxiolytic drug than did mice housed in cages fitted with a nest-box, a running wheel, tubular runways, and nesting material.[17] Infamous experiments involving inescapable electric shock have shown that lack of behavioural control paired with aversive stimuli can produce pathological levels of stress in animals.[18] It has only fairly recently come to light that great apes and elephants are susceptible to post-traumatic stress disorder and other related psychological disorders as defined for human patients by the American Psychiatric Association.[19]

Routine laboratory procedures such as blood collection, gavage (forced feeding), cage moving or cleaning, and handling have all been shown to cause stress in monkeys, rodents, rabbits, and birds, as measured by significant elevations in blood levels of the "stress hormone" corticosterone.[20] Rats show similar responses when other rats are killed in the same room, and mice become increasingly sensitive to painful episodes when they have seen another, familiar mouse writhing in pain.[21]

Evidence for Animal Pleasure

The arguments presented in the preceding section predict that we should find many examples of animals' experience of pleasure in nature. Because there has been very little scholarly attention given to the phenomenon of animal pleasure, most examples remain anecdotal. The most familiar are people's interactions with companion animals. The excitement shown by a dog for an anticipated walk, or a cat's solicitation for another belly rub—these are just two of many ways that our "pets" inform us that they enjoy something and seek more of it. However, while they are scarce, there have been a few very good scientific evaluations of animal pleasure. I will summarise these here.

---

14 Hatch et al., "Longterm Isolation Stress," 507.
15 Sharp et al., "Stress-Like Responses," 8–14.
16 Van Loo et al., "Do Male Mice Prefer," 91–103.
17 Sherwin and Olsson, "Housing Conditions," 33–9.
18 Selye, *Stress Without Distress*.
19 Brüne-Cohrs, Brüne-Cohrs, and McGrew, "Psychiatric Treatment for Apes?" 807; Bradshaw, *Elephants on the Edge*.
20 Balcombe, Barnard, and Sandusky, "Laboratory Routines," 42–51.
21 Langford et al., "Social Modulation of Pain," 1967–70.

Play is perhaps the least controversial of all animal behaviours that may be interpreted as pleasurable. While play has clear adaptive value to the player—for example, it develops physical strength and coordination, and it hones social skills—animals (and for the most part humans) do not consciously play for ultimate, adaptive reasons; they play because it is fun to do so. A series of studies by the American neuroscientist Jaak Panksepp and his colleagues have been directed at rats' affective experience of play behaviour. Juvenile rats spend considerable amounts of their time engaging in rough-and-tumble play. Two of the most common manoeuvres performed by playing rats are 1) for one to playfully bite the nape of the other, and 2) for one to pin the other on the back and playfully "attack" the other's belly. By imitating with the hand these interactions, Panksepp's team was able to measure young rats' motivation and response toward these behaviours. Their techniques involved either gently stroking a rat on the nape, or using vigorous, rapid, scaled-down movements of the fingers of one hand to tickle a rat's belly after flipping him onto his back.

In a typical experimental design, a group of juvenile rats was divided into two groups; members of one group received petting on the nape for a brief period on five consecutive days, and the other group were flipped over and tickled on the belly during the same period. During these sessions, the rats' vocalisations were recorded to count ultrasonic chirps, which rats have been shown to make during presumably pleasurable situations associated with food and mating. Belly-tickled rats made seven times more of these chirps than did petted rats. The difference also increased over the five-day period, suggesting the rats' enthusiasm for being tickled grew. Following a few days' rest, the rats were placed in a chamber, into which a familiar hand was introduced. Both petted and tickled rats approached the hand, but rats accustomed to being tickled ran to the hand four times as quickly as did petted rats. Rats presented with two bars also pressed the tickle bar repeatedly, and the other, passive bar, almost never. The researchers concluded that while both petting and tickling are positive, rats show a strong preference for being tickled on the belly.[22] For all intents and purposes, theirs is a mirthful response, and Panksepp has likened it to the laughter of tickled humans and other apes.[23]

Studies by the Canadian physiologist Michel Cabanac have shown that nonhumans, like humans, will make sacrifices to obtain highly desirable food items. Rats invariably shunned laboratory chow and ran into a cold environment to consume highly palatable foods (meat paté, shortbread, and CocaCola®), and the animals' individual preferences (which varied considerably) were reflected by the amount eaten, the number of excursions, and the time spent feeding in the cold.[24] More recent studies by Cabanac have shown that reptiles are capable of the same nuanced responses to food rewards. A study of juvenile green iguanas showed that these reptiles would trade off the palatability of a bait (lettuce) with the disadvantage

---

22 Burgdorf and Panksepp, "Tickling Induces Reward," 167–73.
23 Panksepp and Burgdorf, "'Laughing' Rats," 533–47.
24 Cabanac and Johnson, "Analysis of a Conflict," 249–53.

of having to venture into a very cold area to retrieve it. As the ambient temperature at the bait decreased, the lizards visited the bait less frequently and for shorter periods, choosing instead to stay under the heat lamp where nutritionally complete (but apparently less rewarding) reptile chow was freely available. Moreover, time interval between sessions with bait (ranging from one to eight days) had no effect on the duration of stay on the bait, suggesting that the lettuce was more of a luxury rather than an indispensable nutritional food source.[25]

While touch is not indispensable to survival in the manner of food, physical contact nevertheless plays an important role in the interactions of many social species. Some primate species spend as much as 20 percent of their waking time engaging in grooming behaviour, and the activity is accompanied by the release of pain-relieving endorphins.[26] Horses also regularly engage in mutual grooming by licking and gently nibbling one another's neck and withers—spots where horses are unable to self-groom. When human researchers experimentally groomed semi-tame Camargue horses, the animals' heart rates dropped significantly, but only when the touch was directed at those areas of the neck that are the preferred grooming sites in this species.[27]

Cleaner-client mutualisms among fishes constitute one of the most intriguing examples of what may be interpreted as a pleasure-mediated animal behaviour. Recorded from both freshwater and marine environments, the most closely studied of these are the relationships between a small, slender, striped reef fish, the cleaner wrasse, and the variety of other reef fish which come for the wrasses' cleaning services.

Cleaner fish nibble loose skin, fungal growths, and fish lice from other fish "clients." They may also pluck necrotic tissue from a client's wounds, which may relieve infection and speed healing. It is a mutualism: cleaners benefit by getting food—delivered buffet-style by clients who line-up patiently to await their turn—and clients get a body-cleansing service. Different species of client are cleaned in a highly specific manner by cleaner fish (and shrimps), who advertise their services with brightly coloured uniforms and perform bobbing/fussing movements to signal their willingness to attend to clients. Clients also may signal their readiness, for instance by orienting themselves vertically in the water, and opening their mouths and gills at appropriate times to allow access to the cleaners.[28]

Several features of the cleaner-client relationship suggest that positive feelings are involved. Invitation postures indicate that cleaners may be anticipating the attentions of clients, and clients the services of their hosts.[29] Tactile stimulation appears to be an important motivator for the interaction. Based on 112 hours of observation of 12 different cleaners, it appears that cleaners are able to alter client

---

25   Balaskó and Cabanac, "Behavior of Juvenile Lizards," 257–62.

26   Keverne, "Primate Social Relationships," 1–36.

27   Feh and Mazières, "Grooming," 1191–4.

28   Potts, "The Ethology of *Labroides dimidiatus* on Aldabra," 250–291.

29   Ibid.

decisions over how long to stay for an inspection, and to stop clients from fleeing or responding aggressively to a cleaner bite that made them jolt.[30] Cleaners and clients also appear to recognise each other,[31] and they return to their favoured business partner,[32] much as we return to a favourite barber or hairdresser. Finally, cleaners sometime cheat by nipping off some of the client's own skin; established clients behave as if taking serious offence at this, chasing the cleaner around, and/ or shunning the cleaner's future solicitations. This punishment helps to stabilise the relationship between cleaners and their clients.[33]

A similar arrangement has been documented in Kenya's freshwater Mzina Springs, where hippopotamuses come to rest and cool off during the daytime. Fishes of various species gravitate to the hulking mammals, nibbling and plucking at their skin. Observations by biologists/photographers Mark Deeble and Victoria Stone indicate that the hippos are far from passive participants in these cleaning services. They actively splay their toes, spread their legs, and hold open their mouths to provide easy access for the fishes, and they appear to solicit cleanings by visiting "cleaning stations" where fish congregate.[34]

As if not to be outdone by their hippo cousins, warthogs at Uganda's Queen Elizabeth National Park flop down on their sides to be swarmed by mongooses. The little predators swarm over the recumbent warthog, scouring the skin, perhaps for salt or for external parasites. The warthogs show every sign of thoroughly enjoying the attention. They stretch out with their eyes closed, like pampered spa-goers.

These few examples offer a glimpse into the pleasurable aspects of animal experience. Because few scientists are examining the experience and expression of positive affect in nonhumans, a glimpse is all that is currently available. It is notable that these examples span the spectrum of vertebrates, from those commonly presumed to be the least sentient (fishes) to the mammals. It may be argued whether or not invertebrates can feel pleasure, but there ought to be little doubt that all animals with backbones can.

Pleasure's Moral Worth

The capacity for pleasurable experience has enormous moral implications. Sentience, which is the foundation of ethics, comprises a continuum of physical and emotional experience ranging from the depths of agony at one extreme to the heights of ecstasy at the other. It follows that pleasure roughly makes up half of the spectrum of sentient experience. Utilitarianism, which was first expounded

---

30   Bshary and Würth, "Cleaner Fish," 1495–501.
31   Tebbich, Bshary, and Grutter, "Cleaner Fish," 139–45.
32   Bshary and Shäffer, "Choosy Reef Fish," 557–64.
33   Bshary and Grutter, "Punishment and Partner," 396–9.
34   Deeble and Stone, "Kenya's Mzima Spring," 32–47.

as a formal philosophy by Jeremy Bentham (1748–1832), favours actions that optimise pleasurable outcomes while minimising negative ones. Notably, Bentham regarded animals as serious objects of moral concern, based on their capacity for both pain and pleasure.[35] That Bentham's utilitarianism has also been called Hedonistic Utilitarianism further acknowledges the importance of pleasures to considerations of right and wrong. Today, utilitarianism remains a prominent philosophical doctrine, and its most influential champion, Peter Singer, has argued strenuously for the inclusion of animals in the calculus of moral consequences.[36]

If an individual's sensory experiences were confined only to the negative realm—that is, if that individual could only feel pain or non-pain—then it can be seen that that individual has interests. Specifically, the individual seeks to avoid pain. However, for an individual whose sensory capacities include pleasurable experiences, that individual's interests are broadened to include not just the avoidance of pain, but the obtaining of pleasure. Until there is any compelling evidence to the contrary, we may fairly confidently assume that most organisms equipped to feel pain are also probably equipped to feel pleasure. It follows, then, that for organisms which qualify as sentient, life takes on a profound new quality: intrinsic value.[37] Intrinsic value means that life is of value to the individual who possesses it, independent of any value that the individual may have for another, for example as a source of food or revenue. Depending on the relative needs of individuals with conflicting interests, it may be argued whether or not the interests of one individual trump those of another; but at the very least, a life of intrinsic value carries some moral weight.

It is worth pointing out that humans have not necessarily set the standards for sentience. It is far from clear that the size and sophistication of an animal's brain is proportional to that animals' sensitivity to pains or pleasures. That a less intelligent animal may be no less sentient than a more intelligent one is not a new idea. The British Clergyman Humphrey Primatt (1735–1779) eloquently expressed the tenuousness of linking suffering to intellect: "Superiority of rank or station exempts no creature from the sensibility of pain, nor does inferiority render the feelings thereof the less exquisite."[38] Contemporary neuroscientists and ethologists have recently suggested that other animals may experience some feelings more intensely than we do.[39] As far as is known from physiological studies, the perception of noxious stimuli and their conduction to parts of the brain that register pain are fairly identical processes among the different mammals that have been so examined.[40] Because many animals have more acute senses then we do, they may feel certain things more intensely than we do. Recognising a

---

[35]   Bentham, *Introduction to Morals*.

[36]   Singer, *Animal Liberation*.

[37]   For a discussion of this term see Regan, *Case for Animal Rights*, 142.

[38]   Primatt, *Dissertation on Duty*, 7.

[39]   Burgdorf and Panksepp, "Neurobiology of Emotions," 87–98.

[40]   Loeffler, "Pain and Suffering in Animals," 257–61.

painful thing and trying to escape from it should not be any less compelling an evolutionary imperative for a rodent than for a primate. Thus, being pierced with a needle need not feel less painful to a mouse than to a man. In some situations, the fact that we know we are about to be jabbed with a needle, and why, may lessen (or in other cases increase) the pain of the event for us.

How shall we apply the implications of lives worth living to moral decision-making? Specifically, how does the reality of animal pleasure inform the human-animal relationship? Throughout recorded history and into the current era, animals' interests have fairly routinely and consistently taken a back seat to those of humans. We continue to treat animals according to a doctrine of "might-makes-right," or "bright-makes-right." Animals are still legally defined as the property of humans. We have some laws designed to protect animals, but almost all of these are aimed at improving welfare within the paradigm of human use, rather than securing rights or freedoms that are universally ascribed to humans (the PETS act, which Marsha Baum describes later in this volume, is a case in point). Paradoxically, while our knowledge of and moral concern for animals are at their most developed in our history, we continue to harm and kill animals at unprecedented rates,[41] due mainly to an increase in the number of human mouths to feed, and the global spread of Western, meat-centered dietary habits.

In sum, we have subjugated nonhumans to our needs because we have the power to do so. And it is only relatively recently in our moral evolution that we have begun to ask whether we should.

Broadly, pleasure's moral significance informs two aspects of our relationship to other animals. Both pertain to the denial of pleasures. First, when we keep animals in impoverished conditions, as we do in intensive confinement farming systems, laboratory cages, and many zoos, we deny them the opportunity to express natural behaviours and activities from which they may derive pleasure. Studies of rats and mice show that they retain the behaviours of their wild ancestors despite being bred in captivity for many generations;[42] these behaviours include foraging, locomoting, burrowing, exploring, and choosing specific social partners. It is also well documented that, given the opportunity, rodents are highly motivated to live in more complex surroundings, with the company of conspecifics, where they can engage in these behaviours.[43]

Some scholars object to this denial-of-pleasure argument with the claim that captivity offers the animal a safe refuge from the vicissitudes of life in the wild.[44] Wild existence can be hazardous, and some species may live longer in captivity

---

[41]   Animal Studies Group, *Killing Animals*, 1–4.

[42]   Estep, Lanier, and Dewsbury, "Copulatory Behaviour," 329–36; Boice, "Burrows," 649–61.

[43]   Dawkins, "Behavioural Deprivation," 209–25; Balcombe, "Laboratory Environments," 217–35.

[44]   Tannenbaum, "Paradigm Shift," 93–130; Blanchard, "Animal Welfare," 89–95.

(though others, such as cetaceans and elephants, do not).[45] The problem with this objection is that a safer life is by no means a better life. We would be safer if we stayed home, never traveled abroad, and never ate food prepared by others, for instance, but few of us would choose this mundane sort of existence. Another problem is that animals—like humans—prefer not to be confined; it is fairly axiomatic that animals prefer freedom to being in a cage.

The second form of denial of pleasure pertains to death. In killing animals— especially young, healthy animals—we cause harm by denying them the opportunity to experience rewards that life would otherwise offer them. It may be claimed that a dead animal misses nothing. But the main reason that our criminal system treats murder so seriously is not that the victim may suffer—though that certainly compounds the crime. Murder is wrong because life, specifically that portion of life yet to be experienced, has value. Thus, killing is the greatest harm that can be done to conscious, autonomous beings,[46] and pleasure is firmly rooted in the harm committed.

Conclusion

The human-animal relationship is an evolving one. There are many signs that we are giving animals more consideration than in any prior era. A century ago there were not enough books on animal ethics to fill a desk drawer. Today there are enough to fill a small library. There are at least 12 academic journals in the field of animal ethics, and currently more than 200 courses taught at American and Canadian universities. Animal law is a thriving discipline, with 78 courses currently offered at American universities and colleges, according to the Animals and Society Institute. The Humane Society of the United States reports a record 91 animal protection laws passed in the United States in 2008, eclipsing the previous record of 86 new laws in 2007. On both sides of the Atlantic, legislation is being enacted to eradicate the worst of factory farming practices; in 2008, California voters overwhelmingly elected to support a ban on calf and pig crates and batter cages. The use of chimpanzees in laboratory research is now banned in several countries.

These advances are all informed by a growing awareness that animals matter. They matter for two reasons. First, they matter because they are capable of suffering, and we recognise that their suffering is intrinsically bad. Second, they matter because they are capable of pleasure, which means they have lives worth living. By considering animals, we do both them and ourselves a service. We acknowledge animals as sentient in the fullest sense—that is, as possessors of a quality of life. If we view animals' interests solely in terms of avoiding pain and suffering, then the case for their moral protection appears sound. When we include their capacity for pleasure, the case is made stronger.

---

45   Clubb et al., "Compromised Survivorship," 1949.
46   Balluch, "Animals Have a Right," 281–6.

Chapter 7
The Nature of the Experimental Animal: Evolution, Vivisection, and the Victorian Environment

Jed Mayer

The languages of environmental advocacy and of animal welfare developed cooperatively over the course of the nineteenth century, negotiating the often conflicting demands of romantic sympathy and rational science. Although conflicts would later arise between the practices of nature conservancy and resource management and the principles of animal advocacy and protection, defences of animals and their environments enjoyed a momentary alliance. In the later nineteenth century, particularly in Great Britain, criticisms of industrial pollution, deforestation, and urban sprawl drew liberally from the rhetoric and imagery developed through debates over the issue of animal experimentation. As citizens of the first industrial nation, British social critics had expressed their concerns over the country's dwindling green spaces for the better part of a century, yet it was the vivisection debates of the 1870s and 1880s which decisively transformed these scattered criticisms of industrialisation into a coherent and concerted challenge to the ethical limitations of human dominion over the nonhuman. The British antivivisection movement developed a rhetoric and a vocabulary for challenging the anthropocentric basis of unchecked scientific and industrial development, and helped to unite disparate protest movements under the principles of interspecies sympathy and biological kinship.

The nineteenth-century shift in sensibility towards the nonhuman world, charted by Keith Thomas, Barbara T. Gates, Harriet Ritvo, and others, produced an image of nature in which nonhuman animals, as well as the trees, rivers, and rocks they lived among, were regarded as having intrinsic value, and not merely use-value as natural resources.[1] Growing awareness of the vulnerability of organic

---

[1]   Thomas, *Man and the Natural World: Changing Attitudes in England, 1500–1800.* Despite the time span indicated by its subtitle, this book draws most of its evidence for a fundamental shift in British sensibility regarding the nonhuman from the late eighteenth through the nineteenth centuries. Harriet Ritvo's writings, particularly *The Animal Estate*, have focused primarily on nineteenth-century cultural attitudes towards the animal, but her more recent work has come to focus on nineteenth-century green movements, as seen in her recent essay "Manchester v. Thirlmere," 457–81. In her work on the role of Victorian and Edwardian women in nature conservation and animal advocacy, Barbara T. Gates has

environments to industrial development, and of animals to human exploitation, produced a widespread recognition of human responsibility towards the nonhuman. Leslie Stephen was one among many public intellectuals who would applaud this shift in nineteenth-century sensibility: "The widening recognition of the truth that men are responsible for the happiness of the creatures whose lives are entirely dependent upon us in every stage of their existence ... shows a gratifying tendency, which clearly ought to be encouraged."[2] Yet this "gratifying tendency" seemed implicitly to be threatened by recent trends in physiological research. Since, according to Stephen, "the most distinct advance in morality has been the growth of tender feelings towards men and animals," the place of the nonhuman animal in scientific research would necessarily become an unusually vexed one, making more apparent the growing divide between otherwise parallel advances in scientific knowledge and the progressive culture of compassion. The most dramatic and generative public debates over the limits of human dominance in the nineteenth century were waged, not at the borders of ancient forests or at the gates of smoke-belching factories, but around the physiological laboratories which began to proliferate in British research institutions in the 1870s.

Before 1870 vivisection was rarely practiced in Great Britain. Traditionally associated with the perceived materialist excesses of Continental scientific inquiry, the practice was widely regarded by mainstream British medical and scientific communities as cruelly in excess of its possible benefits. With the increased acceptance of the Darwinian theory of evolution, however, the relatively moribund field of comparative anatomy acquired a renewed sense of purpose. Figures like T. H. Huxley, concerned with the professional advancement of British science, dedicated themselves to bringing the experimental regimen of Continental physiology to native shores. As Hilda Kean observes: "The real growth of vivisection in Britain dated from Darwin's arguing for an understanding of the commonality between species; it also dated from the dissemination of Claude Bernard's pioneering work on physiology within the scientific community."[3] By importing Continental laboratory practices into British institutions, upwardly mobile British physiologists sought increased public and professional recognition for the "new biology," but this recognition also ran afoul of the century's progressive culture of compassion. Moreover, the high social and professional profile cultivated by the new men of science linked them to the captains of industry who were increasingly the subject of public criticism for the pollution and waste produced by their commercial

---

shown many significant connections between forms of activism traditionally regarded as separate. For the romantic context, see Perkins's *Romanticism and Animal Rights* and Kenyon-Jones's *Kindred Brutes*.

2   Stephen, "Thoughts of an Outsider," 477.

3   Kean, *Animal Rights*, 97. For more on the influence of evolutionary theories on the comparative dimension of Victorian physiology, see Geison, *Michael Foster*.

operations.[4] The wealth of knowledge obtained in the physiological laboratory came to be perceived as entailing a moral and ecological cost comparable to that of other innovations associated with the industrial and scientific revolutions.

Far from distancing their work from the natural world outside of the laboratory, the defenders of vivisection frequently emphasised the connection between the experimental animal and the larger world of nature. In Claude Bernard's vivid words, the physiologist "stir[s] the fetid or throbbing ground of life" in the experimental laboratory.[5] The scientist is like an explorer in that "true science is like a flowering and delectable plateau which can be attained only after climbing craggy steeps and scratching one's legs against branches and brushwood." In *The Fortnightly Review*, John Bridges described the vivisectionist as a kind of "scientific geographer at home" obtaining "precise knowledge of the geographical structure of a country, of the elevation of its plateaux and mountain ranges, of the geological features, and of the average rainfall in different latitudes and longitudes."[6] The country described in this rather improbably extended metaphor is, in fact, the inner landscape of the experimental animal, an internal geography which advocates of vivisection in the nineteenth century frequently represented as coterminous with the landscapes which had long been the province of natural historians. Yet of course these landscapes, and the animals who inhabited them, had a growing number of defenders and guardians; thus, the rhetorical expansion of the physiologist's dominion outside of the laboratory came into conflict with the evolving principles of environmental advocacy as well as the antivivisection movement.[7]

The ideological differences between the defenders of vivisection and their opponents were especially pronounced in their alternative responses to the implications of Darwinian theory. Since the evolutionary connection between humans and nonhuman animals had served to underwrite what one physiologist enthusiastically described as the "renaissance period of English physiology," many antivivisectionists, like the prominent Frances Power Cobbe, rejected the principles of evolution outright, drawing a connection between the allegedly dehumanising implications of evolution with the imputed brutality of vivisectionists.[8] Cobbe

---

[4]   In his study of the professional politics of scientific research at Cambridge, John Pickstone characterises Michael Foster and other associates of Huxley's as highly ambitious "exponents of the new analytical sciences, and they wanted jobs for themselves and others, especially as teachers." Pickstone, "Science in Nineteenth-Century England," 58. At the International Medical Congress, held in London in 1881, Michael Foster decried the effects of the antivivisection movement on the professional possibilities of physiological experimentation: "At the present day careers are opening up, and a fair amount of useful work is, I trust, being done, or rather perhaps would be done, had not, in this country, physiology fallen upon evil days of a kind unknown in the eighteenth or any other century." Wilks, "Vivisection," 942.

[5]   Bernard, *Introduction*, 15.

[6]   Bridges, "Harvey and Vivisection," 14.

[7]   For a more detailed account of the rise of Victorian environmentalism, see Ranlett, "Checking Nature's Desecration," 198–222.

[8]   Rutherford, "Opening Address," 457.

and other antivivisectionists also perceived a strong connection between the Darwinian metaphor of the struggle for existence and the allegedly amoral practices of the experimental physiologists.[9] Such comparisons were, in fact, made available by vivisection advocates themselves, who often represented the physiological laboratory as a place of deadly struggle between the physiologist and his experimental subject. In an impassioned defence of vivisection to readers of *MacMillan's*, Michael Foster, Cambridge's first full-time professor of physiology and its most active promoter, reasons that:

> The success of the human race in the struggle for existence depends on man's being well fed; man is therefore justified in slaying and eating a sheep. The success of the human race in the struggle for existence is dependent on knowledge being increased; man is therefore justified in slaying a frog or a rabbit, if it can be shown that human knowledge is thereby enlarged.[10]

Indeed, man is not only justified in such an act, but is obliged to do so for his survival: "Unless man destroys animals, animals would soon destroy man."[11] Employing Darwin's theory with a literalness that might have surprised its framer, Foster transforms the physiological laboratory into a place of deadly struggle for man's ascendancy over the brute natural order:

> Mr. Darwin has shown that the lives of all living beings are shaped by "the struggle for existence." Man's life is a struggle for existence with his fellow-men, with living animals and plants, and with the lifeless forces of the universe. The very conditions of his existence lay upon him the burden, and in so doing give him the right, to use the world around him, the lives of animals included, to aid him in his strife. Imagine the results of forbidding man to take away the lives of animals.[12]

By pitting humanity against the world of "living animals and plants," Foster and other British champions of vivisection provided early animal rights and environmental advocates with a clearly defined point of connection, facilitating a more widespread opposition to the espoused principles of scientific dominion.[13]

------

[9]   Cobbe's starkest denunciation of evolutionary theory is her essay "The Scientific Spirit of the Age" where she argues: "While Darwinism is in the ascendant, the influence of the doctrine of hereditary conscience is simply deadly," 134. The influence of alternative interpretations of Darwinian theory by her fellow animal advocates seems to have softened her view towards evolution, however, as shown in her later essay "The Ethics of Zoophily."

[10]   Foster, "Vivisection," 369.

[11]   Ibid., 368.

[12]   Ibid.

[13]   For a discussion of these late nineteenth-century organisations, united under a broad and inclusive humanitarian ethos, see Kean, *Animal Rights*, 136–64.

Particularly enabling to the conceptual merging of environmental concerns with animal rights were such defences of vivisection as that made by zoologist Edwin Ray Lankester in the pages of *Nature*, where he argues against humanitarian concerns in favour of a more just recognition of the value of human progress:

> It is futile to bewail "the tremendous cost" at which such progress is made. Nature is inconceivably costly … for no progress is made without endless suffering and immense destruction. Our very dinner-tables reek with the evidences of "the tremendous cost"—the pangs of slaughtered sheep, the groans of over-worked horses, the disfigurement of Nature's sacred face by agriculture—by which our corporeal means of progress is attained. And are we to be so inconsistent as to refuse to undertake the very highest occupation of humanity, the ascertainment of the order of Nature, because it adds to this "cost" of our existence?[14]

Such arguments were often raised by advocates of vivisection as a way of challenging the alleged hypocrisy, or inconsistency, of their opponents. Yet, while it may be argued with some justice that many within the antivivisection movement tended to focus on only this one issue out of many relating to the ethical treatment of animals, the pressing ethical questions raised in regards to experiments on animals came increasingly to be asked of practices extending well outside of the laboratory.

Much of the impetus for extending discussions of animal experimentation into other spheres came from proponents of the new biology. If physician Samuel Wilks could claim in 1881 that it was "rare to find a consistent anti-vivisectionist, who, while opposing experiments on animals, would join in a crusade against all existing cruelties, without reference to the question of utility," in just a few years such crusades would indeed emerge directly out of the vivisection debates.[15] By comparing vivisection to other forms of alleged cruelty to animals, the scientific lobby may have called their opponent's bluff, but they did so at the price of presenting an image of human dominance at its most ruthless. In a monograph published by the Physiological Society, *Physiological Cruelty; or, Fact vs. Fancy*, the pseudonymous author Philanthropos presents an especially grim catalogue of cruelties to animals:

> We kill them (without anaesthetics) not only that we may have food and clothing, but that the food may be varied and attractive, and the clothing rich and beautiful. We subject them to painful mutilations in order to make them more manageable for service, to improve the flavour of their flesh, and even to please our whimsical fancies. We imprison them in cages and Zoological Gardens, to improve our knowledge of Natural History, or merely to amuse ourselves by looking at them. It is abundantly clear that in all our customary dealings with animals we apply to them without scruple the law of sacrifice, and interpret it with a wide latitude in our own favour.[16]

---

14 Lankester, "Vivisection," 145.
15 Wilks, "Vivisection," 938.
16 "Philanthropos," *Physiological Cruelty*, 36.

Such rhetoric would seem to have had an effect opposite to that intended by the nineteenth-century defenders of vivisection, as animal rights advocates expanded their charter well beyond the relatively cautious platform promoted by mainstream animal welfare groups like the moderate RSPCA. Henry S. Salt, author of *Animal Rights* (1892) and founder of the Humanitarian League, stressed the centrality of vivisection as a metaphor for other forms of cruelty. In a letter to the *Daily News*, Salt argued that:

> the vivisection of the laboratory is but one of many forms of torture that are inflicted, quite needlessly, on millions of highly sentient beings in the interest of science, sport, fashion, or appetite. Even the term "vivisection" is not applicable to the scientific practice only. The clumsily butchered ox, the gelded horse, the mangled pheasant or hare, the half-flayed seal—each and all of these are, in the most literal sense of the word, vivisected.[17]

Antivivisection publications initially stressed the importance of addressing animal experimentation as a "first necessity" among animal welfare issues, but this priority would gradually be regarded as paradigmatic of human-animal relations generally. If justifications for using animals in scientific experimentation were founded on the same principles of human dominance as those which authorised the use of animals for food, fashion, and amusement, there were many within antivivisection circles who were prepared to subject those principles to closer scrutiny, along with the practices they justified.

By placing vivisection in the vanguard of humanity's struggle against the nonhuman world, its advocates had provided their antagonists with a vivid image of nature reduced to the logic of the experimental laboratory. Prominent critics of industrialism came to recognise similarities between the language promoting scientific development and that which bolstered industrial and economic speculations, and criticisms of science and industry became more closely entwined. Socialists like William Morris emphasised the role of *laissez faire* capitalism in scientific and industrial development alike. Asking what we might expect of science in providing positive solutions to the problems of industrial pollution and waste, Morris sceptically observes:

> I fear [science] is so much in the pay of ... the counting-house and the drill-sergeant, that she is too busy, and will for the present do nothing. Yet there are matters which I should have thought easy for her; say for example teaching Manchester how to consume its own smoke, or Leeds how to get rid of its superfluous black dye without turning it into the river, which would be as much worth her attention as the production of the heaviest of heavy black silks, or the biggest of useless guns.[18]

---

17   Salt, "Vivisection," 155.

18   Morris, *Collected Works*, 24–5.

Criticisms of industrialism could often merge with criticisms of science in presenting a pathology of human dominance in which vivisection, pollution, and urban sprawl were different symptoms of the same disease. In her essay, "The Ugliness of Modern Life," Victorian woman-of-letters Ouida presents an image of urban decay in which humans are trapped in cages like laboratory animals, victims of failed experiments in city planning:

> It is natural that the people shut up in these structures crave for drink, for nameless vices, for the brothel, the opium den, the cheap eating-house and gaming booth; anything, anywhere, to escape from the monotony which surrounds them and which leaves them no more charm in life than if they were rabbits shut up in a physiologist's experimenting cage, and fed on gin-soaked grains.[19]

Alterations of the physical environment of Victorian England were often presented by social critics as altering the country's moral or spiritual environment. For antivivisectionists like Anna Kingsford, experimentation was more than a metaphor for such alterations in physical and spiritual surroundings: "[W]hen the vivisectors ask us angrily, 'What right have you to meddle with the researches of scientific men?'" Kingsford argues, "we turn upon them with greater anger and retort in our turn, 'What right have you to render earth uninhabitable and life insupportable for men with hearts in their bosoms?'"[20] Such arguments provided a nexus between the century's developing criticism of industrialisation and more recently developed challenges to the principles of human dominance, as profiteering experiments in science and industry came to be seen as rendering "earth uninhabitable and life insupportable."

Notions of physical and emotional kinship emerging out of the vivisection debates came to shape nineteenth-century criticism of industrial and economic exploitation in subtle but significant ways. John Ruskin, whose influence on early environmentalism can hardly be overestimated, based his critique of scientific authority on an alternative notion of dominion which owes much to his participation in the antivivisection movement.[21] In the final volume of *Modern Painters*, Ruskin argues:

> Let [Man] stand in his due relation to other creatures, and to inanimate things— know them all and love them, as made for him, and he for them;—and he becomes himself the greatest and holiest of them. But let him cast off this relation, despise and forget the less creation round him, and instead of being the light of the world, he is a sun in space—a fiery ball, spotted with storm.[22]

---

19 Ouida, "The Ugliness of Modern Life," 219.

20 Kingsford, "Unscientific Science," 305.

21 See Mayer, "Ruskin, Vivisection, and Scientific Knowledge," 200–222.

22 Ruskin, *Works of John Ruskin*, vol. 7, 263.

This view of human-animal kinship provided the conceptual basis for Ruskin's opposition to vivisection, and he came to feel strongly enough about recent developments in scientific research to resign his Slade professorship at Oxford over the endowment of a physiological research laboratory at the University. During a debate on the issue Ruskin argued that the vivisectionist's knife effectively cut "the great link which bound together the whole of creation, from its Maker to the lowest creature."[23] If evolutionary theory had provided British physiologists with a theoretical foundation for their research practices, it had also provided their opponents with scientific support for "the great link" between humans and the nonhuman world which animal welfare and environmental advocates had been elaborating for the better part of a century. Writing as "an evolutionist to evolutionists," art critic and nature writer Vernon Lee, in her 1882 essay on vivisection for the *Contemporary Review*, urges fellow progressives to take heed of the moral lesson of evolution in recognising that vivisection:

> is contrary to the nature of the highest result of our gradual evolution. I mean that by preferring in this case the advantages which our race might gain at the expense of wholesale and profitless agony to another race, we are laying obliterating fingers upon those delicate moral features which have thus slowly and arduously been moulded into shape.[24]

The "evolutional morality" promoted here might be seen as uniting an otherwise loosely affiliated set of political and social activists of the later nineteenth century, including the Humanitarian League, the Fabian Society, the Fellowship of the New Life, the London Abattoir Society, and the Commons Preservation Society, all of whom included prominent members who were active participants in the antivivisection movement.[25] Edward Carpenter, prominent socialist and active member of the Humanitarian League founded by fellow animal rights activist Henry Salt, emphasised the importance of scientific knowledge to humanity's moral progress. Yet, as he argues in an antivivisection pamphlet:

> To sacrifice—in the thirst for some fresh detail of information—whole hecatombs of living creatures ... is to blind ourselves to that greatest and most health-giving of all knowledge—the sense of our common life and unity with all creatures. It is to sacrifice the greater to the less; it is to suffer loss rather than to effect gain.[26]

A central paradox in the development of animal rights and environmental advocacy is that this foundational sense of a "common life and unity with all creatures" was

---

[23]   Ruskin, *Works of John Ruskin*, vol. 34, 643.

[24]   Lee, "Vivisection," 217.

[25]   In his autobiographical *My Days and Dreams*, Carpenter describes the plethora of new societies emerging in the early 1880s, which "marked the coming of a great reaction from the smug commercialism and materialism of the mid-Victorian epoch, and a preparation for the new universe of the twentieth century," 240.

[26]   Carpenter, *Vivisection*, 12.

developed in conjunction with the researches and writings of the very scientific lobby whose work was accused of sundering this sense of kinship. Indeed, the very word "ecology," along with its derivations, emerged in English through the late nineteenth-century writings of experimental physiologists and zoologists. While the first usage in English was in a translation of *The Evolution of Man* (1879) by Ernst Haeckel, the leading proponent of evolutionary biology in Germany, its native usage was extended by figures like John Scott Burdon Sanderson (editor of the first experimental physiology handbook in English) in describing the complex relationship between an organism and its physical surroundings.[27] In his 1893 presidential address to the British Association for the Advancement of Science, Burdon Sanderson elaborates upon Haeckel's definition of "œcology" as the study of "the relations of the animal to its organic as well as to its inorganic environment," and describes its connection to his own area of expertise, physiology:

> Whether with the œcologist we regard the organism in relation to the world, or with the physiologist as a wonderful complex of vital energies, the two branches have this in common, that both studies fix their attention, not on stuffed animals, butterflies in cases, or even microscopical sections of the animal or plant body— all of which relate to the framework of life—but on life itself.[28]

In his expansive view of the life sciences, Burdon Sanderson stresses the importance of regarding these internal and external approaches as intimately related within "the great science of nature," biology.

Burdon Sanderson's emphasis on the dynamic relationship between an organism's physiology and its environment, "not on stuffed animals ... but on life itself," reflects the general trend taken by the life sciences, away from anatomical study and towards physiological study of living organisms.[29] Yet, as Haeckel himself had noted in his initial definition of ecology, "physiology has largely neglected the relation of the organism to the environment, the place each organism takes in the household of nature, in the economy of all nature."[30] This criticism of physiology was made in 1866, however, and reflects Haeckel's own advocacy of Darwin's theories as providing a more expansive framework for the life sciences. As Burdon Sanderson's celebratory address emphasises, the uniquely evolutionary outlook of the "renaissance period of English physiology" had gone some way towards answering Haeckel's criticism, by stressing the relationship between the laboratory animal and its natural environment, and promoting the image of the experimental physiologist as a kind of "scientific geographer at home." The ecological tendencies of British physiology are vividly evoked in a series of pamphlets published by the association for the vivisection advocacy group, the Advancement of Medicine by Research (AAMR). In *Experiments on*

---

27 See Stauffer, "Haeckel, Darwin, and Ecology," 138–45.

28 Burdon Sanderson, "Inaugural Address," 465.

29 For more detailed accounts of this shift, see Geison, *Michael Foster* and Pickstone, *Ways of Knowing*.

30 Stauffer, "Haeckel, Darwin, and Ecology," 141.

Brute Animals, John Cleland romantically describes the operator attending to his work, with "open and earnest eye to wait upon nature, and learn from her what her doings are, how her operations are conducted, what those processes are, the sum of which we refer to under the name of life, and how they are modified by each change of circumstance."[31]

Yet the "common life and unity" revealed in the physiological and ecological connections between humans, animals, and their shared environments did not fundamentally alter the ethical justification for animal experimentation. As Cleland argues, "Whatever may be the pains of physiological experiment ... those experiments are never performed but with the view of obtaining and spreading a more accurate and complete knowledge of the processes of life."[32] Such knowledge had practical as well as philosophical implications. Suggestively likening biological knowledge to industrial innovation, Samuel Wilks defends the proliferation of animal experiments in England by arguing: "Just as the steam engine has been built up step by step through a series of years, so is our knowledge a continuous growth after the same fashion."[33] As with those developments in chemistry and physics which fuelled the industrial revolution, so the insights of experimental physiology must come through painstaking study of nature:

> Whether it be a question of the nature of the rocks beneath us, or the composition of the ocean, or of vegetable life or of animal life, the method of inquiry is the same. The rocks are broken and put in the crucible, the water is submitted to analysis, the plant is dissected, and, in order to ascertain the laws which govern its growth and propagation, experiments are made by grafting and by cross fertilization. In animal life the same method must be adopted to unlock the secrets of nature. The question of the animal being sensitive cannot alter the mode of investigation.[34]

But of course it is the very sensitivity of animals to natural and artificial stimulus which provided the basis for the experimental method in physiological investigation. The "secrets of nature" revealed under the knife were of a unique kind, testifying to the intimate connections between organism and environment. Yet the nature of that sensitivity could be interpreted in vastly different ways by advocates of vivisection and their opponents.

The evolutionary history of the nervous system in vertebrates linked humans and nonhuman animals, and provided antivivisectionists with a physiological foundation for extending sympathy to other creatures. This extension of sympathy would come to include the natural world generally, with the experimental animal providing a crucial emotional link between the human and the nonhuman environment. Just as experimental physiology in Great Britain stressed the biological connections between an animal's inner and outer environments, so

---

[31]   Cleland, *Experiment on Brute Animals*, 3–4.

[32]   Ibid., 6

[33]   Wilks, "Vivisection," 947.

[34]   Ibid.

opponents of vivisection came to stress the connection between sympathy for animals and for their natural environments. The Humanitarian League sponsored a series of "Humane Science Lectures" in 1896, which brought together a variety of animal rights and ecological concerns. In his lecture on "The Need of a Rational and Humane Science," Edward Carpenter criticised the practice of "taking one little portion of Nature and isolating it from the rest, and then describing it exhaustively *as if* it really were so isolated."[35] This kind of specialised study creates a division between humanity and nature where there should be unity: "Man has to find and to feel his true relation to other creatures and to the whole of which he is a part, and he has to use his brain to further this."[36] If British physiology had demonstrated the intimate connection between organisms and their environment, it had failed to emphasise the connection of human beings to this larger ecology. As Carpenter argues in an earlier essay on "The Science of the Future," the laboratory revolution in science reflected a more general tendency of isolating the scientific observer from the life of things:

> Is it not a strange kind of science ... which causes a man to try to bottle the pure atmosphere of heaven and then to shut himself in a gas-reeking, ill-ventilated laboratory while he analyses it; or allows him to vivisect a dog, unconscious that he is blaspheming the pure and holy relations between man and the animals in doing so?[37]

In Carpenter's ecological vision, humans, "pure atmosphere" and animals are all part of a larger system, one in which biological connections entail moral responsibilities. Although such criticisms of contemporary scientific practices and attitudes singled out vivisection for special scorn, the biological conception of a web of life in which all things are interconnected was in many respects advanced by the very experimental biologists whose work they protested.

The expansive ecological vision developed over the course of the late nineteenth century's vivisection debates suggests the need for recognising the central role played by the animal in early environmental advocacy. In her study of animals in romantic period writing, Christine Kenyon-Jones has noted that the role of the animal in early definitions of ecology is a central one, and yet this role has been little studied by historians in the "ecological school of criticism."[38] By recognising the ways in which British physiology extended its conceptual range in describing "the total relations of the animal both in its inorganic and its organic environment," to use Haeckel's phrase, we can better understand how ecology developed into a term describing a particular *relationship to* an environment, as well as the *workings of* that environment. Jonathan Bate observes that, from its very inception, ecology had been "a holistic science, concerned in the largest sense

---

35 Carpenter, "The Need of a Rational and Humane Science," 14.
36 Ibid., 32.
37 Carpenter, "The Science of the Future," 87.
38 Kenyon-Jones, *Kindred Brutes*, 142.

with the relationship between living beings and their environment."[39] Yet Bate notes a long gap between this use of the term and the first use of the term in its more modern—ethically and politically inflected—usage by Ellen Swallow, a late nineteenth-century American environmental advocate.[40] While Bate has done much to show the emergence of ecological and environmental attitudes in the years before such attitudes found their proper name, the relative absence of the animal in his work has left part of Romantic Ecology's story untold. The link between the two usages of the term, Haeckel's and Ellen Swallow's, might be said to have emerged through the antivivisectionists, who found in the experimental animal the moral nexus between humans and their environment. Philip Fisher has described the culture of Victorian sentimentality as an "experimental extension of humanity" to oppressed and marginalised populations such as "prisoners, slaves, madmen, children and animals," groups traditionally deprived of a voice and the subjectivity which such a voice articulates.[41] By speaking for animals in laboratories, antivivisectionists may be seen as engaging in an alternative form of experimentation to those which they protested, extending the bounds of sympathy and kinship in order to reclaim those ways of knowing the animal which had been annexed by scientific inquiry.

In his essay "Vivisection as a Sign of the Times," Lewis Carroll prophesied a time "when the man of science, looking forth over a world which will then own no other sway than his, shall exult in the thought that he has made of this fair green earth, if not a heaven for man, at least a hell for animals."[42] By linking the experimental animal to the natural world, advocates of vivisection had effectively provided "this fair green earth" with a voice, mediated by imaginative writers into a "cry of anguish from the brute creation" that might also articulate the plight of an environment under siege.[43] The suffering of the experimental animal came to be regarded as paradigmatic of the plight of the nonhuman world generally, and its sacrifice upon the altar of science was regarded by many as a cost too high for an increasingly humanitarian civilisation to tolerate. Yet the strong ties that developed between early animal rights and environmental advocates proved increasingly fragile. As the century's disparate environmental advocacy groups came to seek an increased legitimacy in public and political circles, based on the successful model of organisations like the National Trust, ideological differences between animal rights advocates and conservationists became more acute.[44] Yet in the final decades of the nineteenth century a powerful unity of aims and values existed which served as the basis of a criticism of anthropocentrism itself.

---

[39]   Bate, *Romantic Ecology*, 36.

[40]   Clarke, *Ellen Swallow*, 120.

[41]   Fisher, *Hard Facts*, 100.

[42]   Carroll, "Vivisection as a Sign of the Times," 5.

[43]   For a revisionist history of the emergence of early environmentalism through the project of empire, see Grove, *Green Imperialism*.

[44]   For an account of the dissolution of the late Victorian ties between animal rights and environmental advocates, see Gould, *Early Green Politics*.

Chapter 8
"Room on the Ark?":
The Symbolic Nature of U.S. Pet Evacuation
Statutes for Nonhuman Animals

Marsha L. Baum

In reaction to the ineffective efforts to induce people to evacuate without their animals when Hurricanes Katrina and Rita devastated New Orleans, Louisiana, the United States Congress enacted the Pets Evacuation and Transportation Standards Act of 2006 (PETS Act). This act and similar acts passed by various states, while touted as being important to animals, were intended to protect the human companions of household pets and service animals. The language of the statutes and the limited legislative history of the federal act indicate that the law's focus was on the value of human beings rather than the value of the animals themselves. This essay briefly describes the legal status and treatment of companion animals in the United States during disasters, focusing on Hurricane Katrina, and comments upon the impact of statutes such as the PETS Act of 2006 on nonhuman animals.

In September 2005, people watching news reports on the progress of the evacuation of New Orleans were horrified to see images of a small white dog scratching at the side of a bus. The dog had just been dragged from the arms of the boy who had been holding him so the boy could be placed on the bus to leave the city. The boy reportedly screamed, "Snowball! Snowball!" before he became so distraught he vomited. The police officer who pulled the dog from the boy's arms was reported to have said that he "didn't know what would happen to the dog."[1]

Snowball was not the only animal left in New Orleans as humans were evacuated. An estimated 250,000 pets, including cats, dogs, birds, and fish, were stranded in the aftermath of the hurricanes and the flooding.[2] Some people left their animals at home with food and water, expecting to be able to return in a few

---

[1] Foster, "Superdome Evacuations Enter Second Day." Stories in the print media, on broadcasts, and across the Internet followed the tragedy of the boy's loss and offered reports of alleged reunions of Snowball and the boy. No one has yet verified that the boy and the dog were reunited or that the dog survived the separation. Like thousands of other animals, the dog may have been left to wander the streets of New Orleans without food, water, or shelter as human companions were forced to leave the area without him.

[2] "Katrina's Animal Rescue." The loss of animal life from Hurricanes Katrina and Rita and its aftermath is not known. While the cited episode of the television program *Nature* reported 250,000 pets, Representative Christopher Shays, co-sponsor of the PETS Act,

days. Some people who stayed with their animals through the storm were later required to leave, a few allegedly at gunpoint,[3] without their dogs or cats or other companion and service animals.

During the news coverage of the flooding and disorganisation of the evacuations and rescue efforts, the U.S. population saw stranded dogs and cats, messages spray painted on sides of buildings asking rescuers to save animals left behind, and stories of animals lost, shot, or starving in the weeks following the hurricanes. This news coverage and the images imprinted in the minds of the American public led to federal legislation that was claimed to be a preventative measure against such disasters for animals in future weather events.[4] These headlines overstated the impact of the legislation for animals and underemphasised the underlying reasons for the legislation: to save humans, not nonhuman animals.

Animal Companions in the United States

Companion animals play a special role in American homes. In 2005/2006, the population of companion animals included 90.5 million cats, 73.9 million dogs, 16.6 million birds, 11 million reptiles, and 18.2 million small animals such as hamsters and guinea pigs. Two-thirds of U.S. households included a nonhuman animal member.[5] Animals are considered to be children or family members by a large number of their human companions. According to psychological research, "70–90% of pet owners describe their pets as family members."[6]

People who need to evacuate do not want to leave without their entire families. Since a large percentage of nonhuman animals are considered children or family members, people want to be able to take the animals with them in event of evacuation.[7] However, in 2005, nonhuman animals such as Snowball were not allowed in the shelters with people or on the transportation removing people from

---

stated that "estimates are that some 600,000 animals either died or were left without shelter as a result of Hurricane Katrina." Associated Press, "House Passes Pet Evacuation Bill."

[3]   Davis, "Saving New Orleans' Animals."

[4]   Various animal group websites and news headlines described the new federal legislation as requiring evacuation of animals in the event of a disaster. See, e.g., Simmons, "No Pet Left Behind"; Browning, "Pet Evacuation Act Passed by House" ("State and local emergency preparedness operational plans will address the needs of pet owners so that no beloved pet will need to be left behind during a disaster").

[5]   American Pet Products Manufacturers Association, "National Pet Owners Survey."

[6]   Blum and Silver, "Why is it Important."

[7]   In a recent survey of people in high-risk hurricane areas, the Harvard School of Public Health Project on the Public and Biological Security found that 27 per cent of respondents would not evacuate because they did not want to leave their pets. "Hurricane Preparedness."

the area,[8] leaving the humans who wanted to take their companions with them without the means to do so.

Legal Status of Nonhuman Animals in the United States

Despite the prevalence of animals in American families and the bond between humans and nonhuman animals evidenced by people's expressed perceptions that their nonhuman companions are children and family members, animals are chattel under the law of the United States, perhaps different from the sofa or the family car,[9] but simple property nonetheless. They are property in commercial transactions; for example, animals can be merchandise in a contract of sale[10] or can be exempt from creditors as personal property under debtor/creditor law.[11] Remedies for harm to an animal are based upon the dollar amount spent on an animal by the human owner, not upon the pain and suffering of the particular animal.[12] In determining remedies for loss of an animal, even for a companion animal that is considered part of the family, the value of the animal has historically been assessed as the purchase price or market value,[13] just as the value of any other piece of personal property would be assessed.

The predominant legal status of animals in the United States is therefore based on anthropocentric and property-driven principles. The treatment of animals as property under the law denies the animals any inherent or intrinsic[14] value or rights of their own as living beings and considers the animals to have value only in terms of the human owners' interests, whether purely economic or non-economic. When those courts which have more recently considered emotional value or other non-economic value in determining damages for loss of or harm to nonhuman

---

[8]   Policies restricting transportation and sheltering of nonhuman animals with people served two purposes: to meet the requirements of health and hygiene laws and to provide more space for humans. See Centers for Disease Control and Prevention, "Animals in Public Evacuation Centers" and American Red Cross, "Pets."

[9]   See for example, Hessler, "Mediating Animal Law Matters," 26–7 and Wilson, "Catching the Unique Rabbit."

[10]   Uniform Commercial Code § 2–103(1)(k) (2003).

[11]   Uniform Exemptions Act § 8(a)(2) (1976).

[12]   See Root, "'Man's Best Friend.'"

[13]   Byszewski, "Valuing Companion Animals," 217–18.

[14]   The use of the word "intrinsic" in this context is as defined in the *Oxford English Dictionary* and not as used in recent U.S. case law. In 2007, the federal district court in the Western District of Washington, citing Washington state case law, equated "actual" and "instrinsic," stating, "damages are recoverable for the actual or intrinsic value of lost property but not for sentimental value." *Stephens v. Target Corp.*, 482 F. Supp. 2d 1234, 1236 (W.D. Wash. 2007). However, the *OED* defines "intrinsic" as "[b]elonging to the thing in itself, or by its very nature; inherent, essential, proper; 'of its own.'" *Oxford English Dictionary Online*, 2nd ed., s.v. "intrinsic," 3a.

companions have awarded damages at greater than market value,[15] the emotions and pain considered are the human's, not the animal's.[16]

Under the law, property has less value than human life. As discussed above, recovery for loss of an animal will generally be limited to the market value of the animal. While arguably human lives valued in wrongful death actions are also based on "market value" to relatives of the humans whose lives were lost, the factors considered are much broader and the valuations much greater for human lives than for nonhuman animal lives.[17] In the criminal context, the law allows use of deadly force in defence of another person but does not allow deadly force in defence of property,[18] demonstrating that human lives are valued more highly than property. As property, nonhuman animals are valued less highly than humans.

Impact of Animals' Legal Status on their Human Companions during Times of Disaster

The impact on nonhuman animals of their legal status as property and of the value disparity between humans and property was illustrated by the tremendous loss of animal life during Hurricane Katrina. Since property has less value than humans and has no rights, protection of humans will take priority in rescue efforts, evacuation, shelter, and relief funds. As shown by the story of Snowball, the small dog taken from the boy trying to get onto a bus, people were not allowed to bring their animals onto evacuation transportation.[19] Animals were generally not allowed into shelters with their human companions. The American Red Cross specifically prohibited animals in shelters for humans for safety, health, and hygiene reasons and was supported in its decision by the Centers for Disease Control in its reports on potential health risks of housing animals and humans in the same facilities.[20]

---

[15] Byszewski, "Valuing Companion Animals."

[16] For instance in 1980, a New York trial judge, in deciding the damages for the loss of a mixed breed dog with "no ascertainable market value," assessed the damages to be awarded as the "dog's actual value to the owner," which included the protective value of the watchdog and the loss of companionship of this dog that she had had for almost nine years. *Brousseau v. Rosenthal*, 110 Misc. 2d 1054, 1055–6 (Civil Court of the City of New York, New York County, 1980).

[17] See Lawrence, "Primer on Wrongful Death Claims."

[18] See for example, Connecticut General Statutes Annotated § 53a-21 (2007).

[19] The Emergency Preparedness Plan created for New Orleans in 2008 includes evacuation plans that allow pets on transport vehicles. City of New Orleans, "City Assisted Evacuation Plan (CAEP)."

[20] Centers for Disease Control and Prevention, "Animals in Public Evacuation Centers."

Rescue efforts for animals were limited by mandatory evacuation and lack of access to parts of New Orleans.[21]

The general philosophy for rescue and care of household pets and service animals during disasters has been one of personal responsibility. Rescue organisations and government agencies emphasise the need for humans to be prepared to handle disasters that might require evacuation. Proposed advance plans include identification of pet-friendly lodging, health certificates for animals so they can be sheltered or transported out of state, a proper carrier, and sufficient food and water for several days.[22] Implementation of individual plans such as this would be ideal, but the reality for those people who are without resources during disasters may make the personal responsibility approach unworkable for the nonhuman animals at risk.

While the human companions of animals left in New Orleans during Katrina have been faulted by outside observers for leaving their nonhuman animals behind during evacuation, many people did not have a choice. During the disaster the people most affected were those without resources to evacuate themselves. Many people did not have private transportation available to them or have funds to pay for shelter. Using U.S. Census data, the Center on Budget and Policy Priorities estimated that over half of the population with incomes in the poverty range did not have vehicles;[23] without private vehicles, these people had to rely on public evacuation systems, which did not provide means to evacuate nonhuman animals. Shelters for humans would not accept animals. People who are without resources to take care of themselves will have a difficult time meeting a governmental and rescue organisation philosophy that expects animal owners to take personal responsibility for the safety and shelter of their animals.

By contrast, allowing and even assisting people to take their animals with them in a disaster evacuation will lessen the difficulties experienced by humans. Since the animals are considered family members, humans suffer from their loss, whether that loss is the separation during a disaster or the loss resulting from the inability to reunite with the animal due to death of the animal or readoption of the animal by another person after rescue. People suffer from depression and syndromes such as posttraumatic stress disorder from the loss of a pet just as they would from the loss of a family member. Some studies have shown that the psychological pain of loss of an animal can be longer lasting than that of a human companion because the grief of the loss of an animal is not recognised by society. Either the relationship between the human and the nonhuman animal is not recognised as valid or the loss

---

[21] Senate Committee on Homeland Security and Governmental Affairs, *Hurricane Katrina: A Nation Still Unprepared*, 109th Cong., 2d sess., 2006, S. Rep. 109–322, 261.

[22] See American Red Cross, "Pets" and FEMA, "Information for Pet Owners."

[23] Sherman and Shapiro, *Essential Facts*, 2.

is not acknowledged, leaving the human companion without support or resources for handling the grief of the loss.[24]

The failure to recognise the importance of allowing people to take their nonhuman animals with them in an evacuation can risk harm to humans: those who stay, those who leave, and those who attempt to rescue humans and nonhumans still in the disaster area. Those who stay with their animals[25] are at risk of death and injury from the storm and at risk of illness, hunger, dehydration, and other dangers in the storm's aftermath. Those who leave the area without all of their family members may suffer from increased stress and worry regarding those who stay behind to care for the animals. However, the biggest dangers may result if people are required to leave[26] without their animals. Those who try to return to rescue their animals may place themselves and rescue crews in danger or may face arrest for entrance into an unauthorised area. Rescue workers who attempt to save abandoned animals who are difficult to capture are subject to dangerous environments and, potentially, to injury from the animals they are trying to rescue. Animals who are left alone may escape and create problems for or endanger humans as the animals forage for food on the streets. Animals who are left behind may die, resulting in severe health and hygiene problems for people returning to the area.

The Law of Animals in Disasters before the PETS Act

Prior to Hurricane Katrina, most states and the federal government focused concern about animals in disasters on health and hygiene issues affecting humans; generally, the laws regulated proper handling of animal carcasses and biohazards and looked at ways to prevent disease and other harm to humans. In the case of Katrina, while humans were transported over state lines to neighbouring states for refuge from the storm and its aftermath, nonhuman animals were not allowed unfettered transport out of the affected states. For example, states outside the disaster area, such as Massachusetts, declined to accept animals from the disaster area for fear of diseases such as heartworm.[27]

---

[24]   Doka, "Introduction," *Disenfranchised Grief*, 11. See also Corr, "Revisiting the Concept of Disenfranchised Grief," 43.

[25]   A 2005 Zogby International poll of 3,185 adults across the United States found that 49 per cent of all adults polled and 61 per cent of the pet owners in the sample polled would refuse to evacuate if they could not take their pets with them. Zogby International, "Americans."

[26]   The law provides for forced evacuation in times of emergencies. See, e.g., LSA-R.S. 29:724(C)(2) giving the governor of Louisiana the power to "direct and compel the evacuation of all or part of the population from any threatened or stricken area ... "

[27]   Department of Agricultural Resources, Emergency Order 2-AHO-05.

However, even before Hurricane Katrina, some states followed the State Animal Response Teams (SART) model[28] developed in North Carolina for management of nonhuman animals during disasters. The SART model is intended to provide a public/private partnership at the state level and local or county level (Community Animal Response Team or CART) to prepare for and manage animal care during disasters and to handle recovery efforts after a disaster. This partnership presents the opportunity for local (CARTs) and state (SARTs) teams to plan for and to attempt to avoid all animal-related problems of the kind seen in Katrina. For this model to work effectively, the partnership between public agencies and private entities must be well organised and rehearsed so that coordination and leadership networks are maintained during an actual disaster.[29]

Governmental Responses to Katrina

In the aftermath of Hurricane Katrina, both state and federal governments reacted to the public outcry over the treatment of nonhuman animals during the disaster. State legislatures responded by passing laws to require inclusion of animals in state disaster plans.[30] For example, Louisiana quickly amended its disaster act in June 2006. The amendment requires the Governor's Office of Homeland Security and Emergency Preparedness to assist in the formulation of parish emergency operation plans for "humane evacuation, transport, and temporary sheltering of service animals and household pets in times of emergency or disaster." The mandate to the office was to *require* that persons with disabilities who use service animals be evacuated, transported, and sheltered with their service animals.[31] For household pets (defined as "any domesticated cat, dog, and other domesticated animal normally maintained on the property of the owner or person who cares for such domesticated animal"),[32] the law mandated that the office "assist in the identification of shelters and other state facilities designed and equipped to accept and temporarily house household pets and canine search and rescue teams,"[33] "assist in the development of guidelines for such shelters which may include standards or criteria for admission,"[34] and to "[e]nable, *wherever possible*, pet and pet-owner evacuations for disabled, elderly, special needs residents, and all other residents *whenever such evacuations can be accomplished without endangering*

---

[28]   See North Carolina State Animal Response Team, "SART: State Animal Response Teams."

[29]   Burns, "North Carolina Shares a Model."

[30]   Lauraallen, "Are Government Officials Ready to Evacuate and Shelter Animals in Disasters?" Animal Law Coalition, online posting, 29 August 2008, http://www. animallawcoalition.com/animals-and-politics/article/580.

[31]   The Louisiana Homeland Security and Emergency Assistance and Disaster Act, La. Rev. Stat. Anno. § 29:726(E)(20)(a) (West, Westlaw through 2008 Regular Session).

[32]   29:726(E)(20)(c).

[33]   29:726(E)(20)(a)(ii)(aa).

[34]   29:726(E)(20)(a)(ii)(bb).

human life."[35] The act also provided that pets in cages or carriers were to be allowed on "public transportation during an impending disaster, *when doing so does not endanger human life.*"[36] Under this provision, the primary agency is authorised to provide separate transportation for pets not allowed to use public transportation. The office is to assist agencies "in development of plans to address the evacuation, transportation, and other needs of those household pets that are not evacuated or transported" on the public transportation provided for evacuation of humans.[37]

Until the PETS Act was enacted in 2006, the United States had no federal statutes regarding evacuation of companion and service animals. The PETS Act amends the Stafford Disaster Relief and Emergency Assistance Act, which provides for federal government assistance to the states in times of disaster.[38] Under the PETS Act, states will not be eligible for Federal Emergency Management Agency (FEMA) funds for relief efforts unless state and local emergency preparedness plans "take into account" the needs of individuals with household pets and service animals.[39]

Impact of PETS Act and Louisiana Legislation for Animals: Actual or Symbolic?

Legislation can serve multiple purposes, but a particular act will generally have a primary purpose either to change/control specific behaviour or to respond to a public concern. Laws that provide strong incentives or stiff penalties to encourage compliance are likely to change behaviour and have an actual and immediate impact on a particular issue. However, laws that have limited enforcement mechanisms, either in statutory penalties or in allocated resources such as personnel and funding to enforce, or that do not provide effective incentives for action, are symbolic in nature.[40]

---

[35]　29:726(E)(20)(a)(iii)(aa); emphasis added.

[36]　29:726(E)(20)(a)(iv); emphasis added.

[37]　Ibid.

[38]　42 U.S.C. § 5121 (2000).

[39]　Pets Evacuation and Transportation Standards Act of 2006, Pub. L. 109-308, 120 Stat. 1725 (6 October 2006). On 4 October 2006, Congress enacted the Post-Katrina Emergency Management Reform Act of 2006 to provide appropriations for the implementation of disaster planning, including planning for household pets and service animals. Post-Katrina Emergency Management Reform Act of 2006, Pub. L. 109-295, 120 Stat. 1355 (4 October 2006).

[40]　See generally Tushnet and Yackle, "Symbolic Statutes and Real Laws." Tushnet and Yackle discuss three categories of legislation: instrumental, expressive, and symbolic, with symbolic described as legislation which simply makes a statement and does not attempt to change behaviour. While my use of the term "symbolic" is more closely aligned with Tushnet and Yackle's "expressive" category of legislation, which attempts to change values by signalling appropriate behaviour, I decline to use the term "expressive" in order to avoid confusion with other commentators who use the term "symbolic" to describe a combination of Tushnet and Yackle's "expressive" and "symbolic" categories.

While symbolic legislation may raise awareness of a problem and provide a broader base of support for societal change, symbolic laws create behavioural change more slowly than laws intended to control or alter behaviour, if the symbolic legislation changes behaviour at all. The symbolic or "feel good" legislation is generally passed quickly, in response to public or media outcry, to assure the public that the problem is solvable and that action has been taken, which allows legislators to move on to other governmental business.[41] However, symbolic legislation has been cited as a way for Congress to "provide policy direction to states" and as a means of providing "a moral educative function to both the law breaking and law abiding, demonstrating what is acceptable and unacceptable behavior,"[42] potentially resulting in long-term changes in law and behaviour despite the lack of enforcement, incentives, or penalties in the initial legislation itself.

Both the federal PETS Act and state legislation such as Louisiana's changes to its disaster preparedness act which were passed in response to Hurricane Katrina were largely symbolic in their impact on animals. The horrific images and stories of both people and animals harmed by the limited evacuation options during Katrina for people with companion and service animals resulted in a quick legislative response,[43] but the reaction was not one designed to preserve the lives and safety of nonhuman animals. Despite the headlines which heralded protection for animals in disasters under the law, the Congressional debate of the PETS Act and the language of the acts make clear that the intent of the legislation was not to protect animal lives, but to protect human lives. The headlines may have offered solace for those disturbed by the images of animals in danger who now felt that animals would be evacuated and rescued but the laws were passed to ensure that people evacuated by removing an obstacle to the process.

The debates on the floor of the U.S. House of Representative regarding the PETS Act illustrate the intent of the disaster legislation passed regarding animal evacuation and rescue. Remarks by legislators offering assurances that "animals do not go before people" underscores the limited and symbolic nature of the legislation.[44] The goal seems to have been to ensure that we would not have to

---

[41]   Marion, "Symbolic Policies in Clinton's Crime Control Agenda," 67–8.

[42]   Ibid.

[43]   Louisiana passed its legislative changes within months of the disastrous losses from Katrina. Louisiana Senate Bill 607 was introduced in March 2006 and was signed into law by the Governor on 23 June 2006. S.B. 607, 2006 Reg. Sess. (La. 2006). The chronology of the Lousiana legislation can be found at http://www.legis.state.la.us/. According to the National Hurricane Center, Hurricane Katrina made landfall in Louisiana on 29 August 2005. Knabb, Rhome, and Brown, "Tropical Cyclone Report," 3. The PETS Act (H.R. 3858) was introduced in the U.S. House of Representatives by Representative Tom Lantos of California on 22 September 2005, less than one month later. 151 Cong. Rec. H8340-03 (2005); Extension of Remarks, 151 Cong. Rec. E1943-02 (2005); H.R. 3858, 109th Cong. (2005).

[44]   Remarks by Representative Shays, co-sponsor of the PETS Act, 152 Cong. Rec. H2985-01, H2987 (22 May 2006).

watch starving dogs and cats in the aftermath of disasters, we would not have to witness dogs being taken from small children who already had lost everything else they had, and we would not have to be concerned about whether people would leave their flooded or destroyed homes if their animals were not allowed to go with them. The U.S. representative who introduced the legislation stated that the PETS Act was born in the moment he saw Snowball ripped from the arms of the unknown boy.[45] It was not born in the moment of seeing a kitten perched perilously on an electric meter inches above flood water or of hearing about the dozens of dogs shot in a school after their human companions were forced to evacuate at gunpoint or in handcuffs but, instead, reflected concern over "the singularly revealing and tragic experience" of watching the face of a boy who lost everything.[46]

In both the PETS Act and the Louisiana state legislation, the focus remained upon the human "owners" of the animals and the need to protect human life. The language of the federal statute which requires that state and local emergency plans "take into account the needs of individuals with household pets and service animals prior to, during, and following a major disaster or emergency"[47] does not address the needs of the animals. By legislating in terms of the needs of individuals with animals, Congress made clear that animals do not have an inherent value of their own under the new statute. Animals are not offered any special protection by the planning process. The focus is clearly on the humans involved.

The phrase "take into account" is weak and does not require implementation of a plan; it merely requires consideration of animal evacuation in preparing disaster plans. After consideration, a planning group could determine that animal evacuation is not practical or is too costly, but the group will have met the statutory requirement of taking the animals into account in the disaster planning process.

Even if a plan does include provisions for companion animal and service animal evacuation, there is no penalty for failure to implement a plan in a time of disaster. The phrase "take into account" does not place a requirement on the state and local planning agencies to take action in implementation and so does not ensure protection for animals. While the PETS Act makes FEMA funding dependent upon "taking into account" the needs of those individual humans with pets and service animals, the law does not provide any consequences for failure to implement the plans or even to include the animals in their plans in a meaningful way.

The Louisiana state legislation, with a focus on evacuation, rescue, and shelter of animals "wherever possible" and "without endangering human life," also makes clear that the focus is on humans. The Louisiana plan includes the joint pet-owner evacuation whenever the evacuations can be accomplished without danger to human life.[48] The preservation of animals is to aid the animal owner or caretaker,

---

[45]   Representative Lantos, 152 Cong. Rec. H2985-01, H2986 (22 May 2006).

[46]   Ibid.

[47]   42 U.S.C.A. § 5196(e)(4) (West Supp. 2009).

[48]   Service animals are treated differently than household pets under the Louisiana statute. Service animals *must* be evacuated and sheltered with the human companion as required under disabilities laws.

not to preserve the animals. The animals are considered to have no interests of their own. In a balancing of interests in preservation of life, the human life will outweigh the nonhuman life under the language of the state statute.

In addition to the limits on the action required by the legislative language, the categories of animals included also limit protections. Not all nonhuman animals are considered in the legislation enacted. The focus is on service animals and household pets. While "service animals" is generally broadly defined to include any animal trained to assist a person with a disability, "household pets" is a much narrower category of animals.

Since the PETS Act does not define the term "household pets" within its provisions, the determination as to which animals are included was left to administrative resolution. Other federal statutes and regulations define "household pets" for specific purposes; for example, housing regulations define "household pets" to include: "a domesticated animal, such as a dog, cat, bird, rodent (including a rabbit), fish, or turtle, that is traditionally kept in the home for pleasure rather than for commercial purposes. Common household pet does not include reptiles (except turtles) ... This definition does not include animals that are used to assist persons with disabilities."[49] If the housing regulation definition of "household pet" were used for purposes of the PETS Act, it is unclear if animals such as horses, a pet goat, a pot-bellied pig, or a snake would be covered by the provisions of the PETS Act.[50]

However, on 24 October 2007 FEMA made clear that the categories of animals covered by the PETS Act would be more narrowly drawn than the categories included in the housing regulation and that these animals would be excluded from the protections of the PETS Act.[51] For FEMA funding purposes, "household pet" was defined as "[a] domesticated animal, such as a dog, cat, bird, rabbit, rodent, or turtle that is traditionally kept in the home for pleasure rather than for commercial purposes, can travel in commercial carriers, and be housed in temporary facilities. Household pets do not include reptiles (except turtles), amphibians, fish, insects/arachnids, farm animals (including horses), and animals kept for racing purposes."[52]

The language in the acts mentioned above includes: "wherever possible," "without endangering human life," "household pets," and plans that are to "take

---

[49]   24 C.F.R. 5.306 (1).

[50]   This author presented this assertion at a faculty seminar at the University of Tasmania Faculty of Law, Hobart, Tasmania, Australia, in June 2007.

[51]   Arguably, pot-bellied pigs could be included as a household pet if they are considered to be domesticated, traditionally kept in the home for pleasure, can travel in a common carrier, and can be housed in temporary facility, but the determination is not clear from the definition. Horses, goats, and snakes are clearly excluded under the FEMA definition.

[52]   U.S. Department of Homeland Security, Federal Emergency Management Agency, Disaster Assistance Policy 9523.19 "Eligible Costs Related to Pet Evacuations and Sheltering," 24 October 2007.

into account." These terms all place limits on the protections and services to be offered to evacuate animals. Zoo animals are not covered. Livestock is not covered. Companion animals that do not fit within the FEMA definition of "household pets" are not covered. Those animals that are covered are subject to the limitations of available resources; if human life might be affected or those responsible for shelters or evacuation or transportation determine that the evacuation and shelter of the animals that people have brought with them is not possible, the animals and their human companions will have no recourse. The language and scope of coverage demonstrate that the recently enacted legislation regarding evacuation of animals in disasters is symbolic as far as the animals are concerned.

Value and Danger of Symbolic Disaster Legislation for Animals

Although symbolic legislation does not have the effect of controlling or changing behaviour, it can still provide moral education and policy direction. The legislative change to incorporate companion and service animals into disaster planning does give those interested in the welfare of nonhuman animals a seat at the table and an opportunity to offer recommendations in the planning process which will affect animals in future disasters. As Representative Chandler stated in support of the PETS Act, "legislation like this helps increase the awareness of lawmakers and emergency officials to recognise what animal advocates already know, that pets figure strongly in a person's decision to evacuate to safety. And we certainly want to encourage our citizens to do just that."[53] State and local disaster planning processes now include representatives from animal welfare organisations and offer the possibility of more careful thought regarding the needs of the animals and not just the humans affected when there is an inability to safely evacuate with companion and service animals.

While the legislation that was enacted in 2006 offered a public voice for the animals in the planning process, that legislation can present a danger to animals by giving the public the sense that action has been taken and the problem has been solved, removing the likelihood that real action will be taken on behalf of animals at risk.[54] One of the risks of symbolic legislation is that, by enacting laws quickly in response to public reaction, momentum is lost in creating real change.

In addition, after reading the headlines about legislation such as the PETS Act, people might assume that they can take their pets and service animals with them and demand placement at a shelter or demand transportation for their animals as well as for themselves but find that the reality is no different than it was for Snowball and her human companion in New Orleans. Because the Louisiana statute limits services if human lives are at risk or to "whenever [evacuation and shelter are] possible," animals may still be turned away and humans required to leave their

---

[53] 152 Cong. Rec. H2985-01, H2986 (May 22, 2006).
[54] Aviram, "The Placebo Effect of Law."

pets behind. While the Louisiana statute requires sheltering of service animals in compliance with the Americans with Disabilities Act (ADA),[55] no requirement appears in the statute for evacuation and shelter of household pets. So, in the event of a disaster, an animal may be worse off than if the human companion had taken the previous route of leaving food and water for the animal and returning in a few days to their home. If forced to separate while away from home, the animal may be lost, adopted out, or killed with no recourse by the human companion.

Improving the Situation for Animals in Times of Disaster

While the acts that were passed in response to Hurricane Katrina have a largely symbolic impact for animals, they do offer a starting point for increasing the safety of companion and other animals in times of disasters. The acts have had an impact by raising public awareness of the issues surrounding animal evacuation and rescue and the need for advance planning, but the impact could be strengthened through increased financial incentives and consequences.

The federal government is limited in its ability to mandate action by the states but can use funding as an incentive. The U.S. Senate added a provision to the PETS Act to allow federal funding for creation of suitable accommodations for people with pets and service animals, going beyond the planning stage into implementation.[56] In the PETS Act, the Congress tied the planning for individuals with animals to FEMA funding but required only that the states take into account the evacuation needs of humans with household pets and service animals. Like the offering of grant funds for construction of shelters for evacuation of animals, the PETS Act could have based funding not just on submission of plans but also on implementation.

While Congress may arguably be limited in the type of consequences that may be imposed for failure to implement, state governments have more direct options. States can require training, funding, and clear and detailed plans for implementation and incur consequences for failure to act on local and state officers. Louisiana created a mandatory evacuation and shelter provision for service animals. The law could also have created a mandatory provision for evacuation and shelter of household pets.

Symbolic legislation does not provide the consequences needed to change the results. The greatest danger to nonhuman animals is that the enactment of legislation without "teeth" results in complacency among the populace who assume that the legislative act handles the situation. Realising in the middle of a disaster and attempted evacuation that the law has not changed policies and implementation and having no recourse for the failure of personnel to act as the populace believes the law requires will result in loss of animal lives.

---

[55]   42 U.S.C. § 12101 et. seq. (2000).

[56]   Pets Evacuation and Transportation Standards Act of 2006, Pub. L. 109-308, 120 Stat. 1725, sec. 3 (6 October 2006).

Recognition of the intrinsic value of animals and animal self-interest in preservation and survival is needed to give substance to disaster and emergency plans. Legislation that recognises only the human interest in companion animals and addresses the need to evacuate animals only as a means to save humans continues the legal paradigm of animals as property. Without legal recognition of the inherent value of animals and with the current focus on human preservation during disasters, it is essential that laws have consequences for failure to act. However, the existing legal model of animals as property impedes both recognition of intrinsic value of animals and the imposition of consequences for failure to protect animals.

Chapter 9
Making Animals Matter:
Why the Art World Needs to Rethink the
Representation of Animals

Yvette Watt

Introduction

Despite a long history of continuing debate over the nature of human-animal relationships and the ethical and philosophical dimensions of the issue, it is only in recent decades that the "animal question" has become more widely accepted as an important and serious matter in terms of scholarly study. This historical refusal to see animals as worthy subject matter in disciplines within the humanities, law, and sciences extended to the visual arts, where for many centuries animals were relegated to a lowly position within the hierarchy of important subject matter for artists.

The recent growth of interest in the issue of human attitudes toward and treatment of other animal species has been noted by a number of scholars including Peter Singer, who, in his preface to the book *Animal Philosophy* points out that:

> One way of gauging the extent to which this issue has become more prominent
> is to look to Charles Magel's comprehensive bibliography of writings on the
> moral status of animals. Magel found only 94 works on that subject in the first
> 1 970 years of the Christian era, but 240 works in the next 18 years, up to the
> date when he completed his work. The tally now [2004] would probably be in
> the thousands.[1]

In addition, American historian Harriet Ritvo, speaking about historical studies, notes that, "To put it briefly, animals have been edging toward the mainstream. No longer is the mention of an animal-related research topic likely to provoke surprise and amusement, as was the case twenty years ago."[2]

The development of interest in animals and human-animal relationships is also evident in the visual arts, as demonstrated by the number of recent exhibitions that have taken animals and/or human-animal relationships as the key curatorial

---

[1]   Singer, *Animal Philosophy*, xii. Singer is referring to Magel's *Keyguide to Information Sources in Animal Rights*.

[2]   Ritvo, "Animal Planet," 130.

theme.[3] As an artist who has been working with this subject matter for many years, I wholeheartedly welcome the attention that is being paid to the themes that run through my own work. However, while there is much to embrace as positive about these exhibitions, there is some cause for concern about the manner in which animals were often present—and presented by—many of the artists and curators. The nature of how animals are represented in the visual arts will be teased out in this essay through observations about a selection of recent animal-themed exhibitions of contemporary art in relationship to my own artwork. These include the dozen or so exhibitions grouped under the title of Unsettled Boundaries, as part of the 2006 Melbourne International Festival of the Arts (including the major exhibition titled The Idea of the Animal); Becoming Animal at the Massachusetts Museum of Contemporary Art in 2005; Voiceless: I Feel Therefore I Am at Sherman Galleries in 2007; and Eye to Eye at Dubbo Regional Gallery in 2007.[4]

The Fashionable but Marginalised Animal in Recent Contemporary Art

In 2001 Matthew Collings made the following observation:

> Brits are very fond of animals and children. Their exhibitions are now full of animals, usually mutants of some kind, or sexually aroused, or dead—for example, sharks and pigs by Damian Hirst, which symbolise death. And racehorses by Mark Wallinger, symbolising class, but with the front ends different from the back ends—symbolising mutant breeding.[5]

---

[3]   It is also worth noting that there have been dedicated animal issues of art journals, such as *Artlink* 22, no. 1, 2001 and *Photofile* 70, 2007, as well as the launch in September 2006 of a new online journal, *Antennae*, that is dedicated to "Nature in Visual Culture" and has a strong human-animal studies bent (http://www.antennae.org.uk/).

[4]   For more information on The Idea of the Animal see http://www.rmit.edu.au/brow se;ID=cp0iqdn7cpdc1#animal. Information on the Voiceless exhibition, including images of the works, can be found at http://www.shermangalleries.com.au/artists/inartists/mixed. asp?exhibition=156&artist=31. Some other significant recent exhibitions of contemporary art on this theme include Fierce or Friendly at the Tasmanian Museum and Art Gallery in 2007/2008; Kiss of the Beast at the Queensland Art Gallery in 2005; Fierce Friends: Artists and Animals 1750–1900 at the Van Gogh Museum, Amsterdam and Carnegie Museum of Art in Pittsburgh in 2006; The Animal Gaze at various venues in London, 2008 (concurrent with the symposium of the same name); Bête et Hommes at the Grande Halle de la Villette in Paris in 2007/2008; Going Ape: Confronting Animals in Contemporary Art at the Cordova Museum and Sculpture Park, Massachusetts in 2007; Animal Nature at Regina Gouger Miller Gallery, Carnegie Mellon University, Pittsburg, PA, 2005; Animals and Us: The Animal in Contemporary Art at Galerie St. Etienne, New York, 2004; Animals at Haunch of Venison Gallery, London, 2004; and The Human Zoo and A Painted Menagerie: The Animal in Art 1600–1930, Hatton Gallery, Newcastle University, 2003.

[5]   Collings, *Art Crazy Nation*, 6.

This observation is an interesting reflection on the growing prominence of animals in contemporary art. Significantly, he also refers to the matter of *how* animals are used and represented in the name of art, whereby they are often presented as symbols or metaphors for other issues, are shown dead and sometimes even killed especially, or are depicted in some deviant manner.

Collings's observations reflect my concern that, despite a general postmodern avoidance of the animal as pure symbol, animals were nonetheless more often present in the aforementioned exhibitions as generic signifiers for the natural world, rather than as individual, sentient, and self-interested beings. In her essay in the Unsettled Boundaries programme, Jane Scott suggests that the renewed interest by artists in the representation of animals has come about in part from a fear that the human desire to differentiate ourselves from nature has led us down a dangerous path that is resulting in "environmental degradation [such as] pollution, habitat and species loss [and] climate change," and that the artists in the programme "use animal imagery to explore the nature of our nature."[6] In the same programme, Linda Williams proposes a number of reasons for this current curatorial interest in animals and human-animal relations but notes that the result is a deeply held concern about the consequences of a long-held view of human dominance over the natural world and of animals being considered as the "servants of man," the result of which is an impending human-driven loss of diversity.[7] In drawing a parallel between an acknowledgement of climate change and a rethinking of human ideas about animals in his catalogue essay for the exhibition Voiceless: I Feel Therefore I Am, curator Charles Green also implies an environmental imperative in this recurrence of the animal theme in recent exhibitions.[8] Implicit within this concern over a human-caused ecological catastrophe is the issue of what our anthropocentric attitude means for *humans* as much as what it means for animals. This matter is addressed by Josephine Donovan and Carol J. Adams, who note that:

> We could respond that many efforts on behalf of animals will qualitatively improve humans' living conditions as well, which is likely to be the case. But such an argument reduces analysis of interspecies oppression to a human-centered perspective. Yes, in terms of reducing environmental degradation, challenging the mal-distribution of food because of the squandering of food resources in the production of "meat," and preventing human diseases associated with eating animals, such as heart disease and certain forms of cancer, it is true that it is in human's interest to be attentive to and challenge animal exploitation. But these responses concede an insidious anthropocentrism while trying to dislodge it.[9]

While the catalogue essays for Becoming Animal, The Idea of the Animal, and Eye to Eye tend not to concentrate quite so much on the ecological aspects of

---

6 Scott, "Animals—Unsettled Boundaries," 4.
7 Williams, "Turn to the Animal," 6.
8 *Voiceless*, 3.
9 Adams and Donovan, *Animals and Women*, 4.

human-animal relationships, a significant number of the artists whose work has been chosen for these exhibitions appear to be more interested in using animals as a means to think through issues surrounding ideas of nature, or as metaphors, signifiers, or representations of the human or Other, than they are in making work which honours the animals themselves. This is in keeping with the work of the artists in Unsettled Boundaries and Voiceless.

The use of animals to explore a variety of loose ideas of "nature" or as metaphors and signifiers for other subjects is so strongly represented in these exhibitions that there is no space to go into detail here of every artist who utilises animals in their work for this purpose. However, the work of Rachel Berwick in Becoming Animal is an interesting case in point. For Becoming Animal Berwick presented an installation piece based on an individual animal, Lonesome George, who is the last remaining member of his subspecies of Galapagos Island Tortoise.[10] However, despite Lonesome George apparently being the focus of the work, Berwick is actually more concerned with a general concept of extinction and loss. She states in the catalogue that she "ha[s] always been interested in the subject of loss. In my work I focus on how we deal with loss, the desire to recover that which is lost …"[11] The catalogue essay on Berwick's work observes that "in emphasizing animal extinction, Berwick brings forward the loss that accompanies this movement and relates it to other cultural events: the lost languages of the South American Maypure tribe for example."[12] Despite the fact that Berwick presents us with an individual, named animal, the animal as a distinct entity is in the end made subservient to a range of other thematic concerns. I raise Berwick's work as a particular example to highlight that, even on this relatively rare occasion where an artist has depicted an animal as a unique and identifiable individual, the animal himself is ultimately overwhelmed by the artist's use of him as a metaphor.

Of further concern is that a number of the artists trivialise the animals in their work, making use of animals more for sake of amusement than for any serious investigation of the complexities of the animals themselves or of human-animal relationships. Notable examples include Sam Easterson and Michael Oatman in Becoming Animal and Chayni Henry in Voiceless.

Easterson's work involves attaching specifically designed video cameras to a variety of animals, including an armadillo, a tarantula, a wolf, an alligator, a chick, a buffalo, a cow, and also a bumbleweed. The resulting videos, which capture footage from the animals' vantage point, are then exhibited. In the *Becoming*

---

[10]   According to the *Becoming Animal* catalogue, Berwick's installation consists of two 14 x 16 foot sails that are gradually inflated. The viewer walks around these to view two videos of a tortoise. "In the first video, the tortoise, barely moving, stares at the camera. His throat expands and contracts as he breathes. In the second video, the tortoise walks up to the camera and then retracts into his shell. At the moment he retracts, fans start suddenly blowing wind into the sails, exhaling." *Becoming Animal*, 32.

[11]   Ibid., 36.

[12]   Ibid., 32.

Animal catalogue Easterson comments: "I think it's especially funny to give sentience to objects and animals that are often overlooked … Animals often let me see how funny they are." Michael Oatman's works in Becoming Animal are collages made up of images of birds which he depicts holding guns and wearing military headwear. Oatman quips that "Birds are funny and scary to begin with." In Voiceless, Chayni Henry exhibited a small-scale, rather childlike painting of her dog, Panda. At the bottom of the painting the crudely painted text says: "PANDA—My last and best dog. She was put down after supposedly being too sick to live in 2000 R.I.P." Despite the seriousness of the subject matter, Henry notes of her paintings: "My work is very straightforward. I have a particular sense of humour." The content and comments of these three artists highlight a dismissive trivialisation that lends weight to the problematic perception that animals are not really serious subject matter.[13]

It is even arguable whether some of the works in these exhibitions did in fact deal specifically with animals or human-animal relationships other than in a manner so general and expansive as to be almost irrelevant. For example, the performance by Ann-Sofi Siden for Becoming Animal where she covers herself in mud and assumes the persona of Queen of Mud appears to have little to do with animals or human-animal relationships. The real subject matter of this work would seem to be the issues of "surveillance, gender, psychoanalysis and site-specificity"[14] which are stated to run consistently through her work. Additionally, the work by Sam Jinks in Voiceless, which depicts a life-size and lifelike interpretation of the pietà with the figures of Mary and Jesus being replaced by those of a middle-aged and a very old man appears to have essentially little to do with the exhibition's objective to "reflect on the connections between ourselves and other species in light of the widespread human use of animals and the cruelty surrounding our dealings with them."[15] While curator Charles Green acknowledges that this work "contains no animal imagery," his suggestion that the animal's presence is implied by the commonality of the human and animal bodies in death as meat is unconvincing.[16] One of the consequences of disregarding the animal as an individual agent is a lack of concern for the ethics of our relationships with animals. It is thus notable that aside from Angela Singer's work in Idea of the Animal, Barbara Dover's work in the exhibitions Idea of the Animal and Eye to Eye, my work in the exhibition Eye to Eye, and to some degree Patricia Piccinnini's work in Becoming Animal, remarkably few of the artists who were included in these exhibitions appear to

---

[13] See *Becoming Animal*, 58, 90; *Voiceless*, 8. I should point out that humour is also an important aspect of some of my own work; however, the humour I use operates in a way that does not trivialise the animals, which I will explain in the next section.

[14] *Becoming Animal*, 106.

[15] Sherman Galleries, website, http://www.shermangalleries.com.au/exhibitions/voiceless.asp

[16] *Voiceless*, 3.

take any clear ethical position regarding the way humans think about and treat nonhuman animals.[17]

The position taken by Kathy High (Becoming Animal), for example, highlights the inconsistencies in attitudes towards animals that are present not only in the visual arts, but in society as a whole. For her work titled *Embracing Animal,* High used homeopathic medicines to treat two transgenic rats she bought online. She gave these rats names, Echo and Flowers, and describes them as "my sisters, my twins."[18] However, High also states that she "understand[s] the need to conduct medical experiments on animals," stating "I am not an 'animal rights' advocate."[19] Elsewhere she speaks of an "ancient contract" we have with animals to give them decent living conditions when they "serve us." But she also suggests the need for a "deep intuitive understanding of human and non-human."[20] It is difficult to reconcile High's relationship with Echo and Flowers as "sisters" or "twins" with her belief in the legitimacy of their fate, and that of millions of their kind, as experimental animals whose role it is to "serve" humans. However, further reading indicates that the real agenda in High's work is her own body. High suffers from a chronic disease, and transgenic rats are genetically altered to emulate human diseases. As such, it would appear that Echo and Flowers are in fact substitutes for High.

This use of animals to represent someone or something else is addressed by British-born American artist, Sue Coe[21] who, according to Steve Baker "object[s] strongly to the idea of using animals as symbols, because by using an animal or its (image) as a symbol of or for something else, that animal is effectively robbed of its identity, and its interests will thus almost inevitably be overlooked."[22]

Coe's view goes to the heart of the matter: animals are too often used to stand in for other thematic concerns in recent contemporary art, even when they appear at first to be the primary subject. As Desmond Morris points out, "symbols die hard." He notes that "we still call someone a stupid pig even though we know how remarkably intelligent pigs are when tested scientifically. And we still accept the bald eagle as a suitable symbol of the United States even though ... it is in reality, not a proud hunter but a messy scavenger."[23]

---

[17]   See Baker's essay in this volume for a discussion of Singer's work. Dover is an Australian artist, curator, and scholar with a background in animal rights advocacy. Her work across these areas is heavily informed by her activist background. In the *Becoming Animal* exhibition catalogue Piccinnini comments that she uses her work to "explore ethical issues that are important to our time." However, I have not been able to find any evidence of Piccinnini's personal position regarding the status of animals. See *Becoming Animal*, 104.

[18]   Ibid., 67.

[19]   Ibid. High is making reference here also to Donna Haraway's description of OncoMouse™ as her "sibling."

[20]   Ibid.

[21]   Coe's work is discussed by Baker in this volume.

[22]   Baker, "'You Kill Things,'" 78.

[23]   Morris, "Foreword," 10.

Ultimately, in being made to stand in for something or someone else, the animals are marginalised. This is not a judgement on the success or failure of the artworks on the artists' own terms. Nor do I mean to wholly disregard the value of these exhibitions, which contribute to the current rethinking of human-animal relationships that is taking place across a range of disciplines and within society in general. It is more, I would suggest, a result of what can be at times a simplistic or overly general curatorial agenda which can require nothing more than an animal's presence in an artwork for it to be deemed to fit within the curatorial premise, and a dearth of artists who are truly concerned with the issues surrounding animals as sentient, self-interested, identifiable individuals.

This avoidance of the politics surrounding animal representation is at odds with a rethinking of human-animal relationships in other disciplines where there is an increasing emphasis on the importance of foregrounding the ethical and political issues of human-animal relationships. As Kay Anderson points out:

> The human-animal divide is increasingly being problematized in the human
> sciences, along with other conceptual distinctions of mind-body/male-female
> that over time have interacted with it. Such dualistic thought is under challenge
> by postcolonial and feminist scholars ... The study of animals has thus been
> brought into a culture/society framework from which it has long been excluded
> ....[24]

A consequence of artists avoiding an engagement with the politics and ethics of human-animal relationships is that it is rare to find artists who are prepared to actually take a stance in the work—to make work that openly expresses their own views. The reasons for artists' reluctance to make their socio-political views clear in their work include a concern that such work may be seen as too closed, too direct, or too didactic, and a persistent attitude amongst artists, curators, and critics that art and socio-political issues do not mix, or at least rarely mix well. This is exemplified by Lisa Roet, whose work was included in The Idea of the Animal and Eye to Eye, who has stated that "I have my own views on the political aspect of the ape issue, but prefer to keep them separate. Art is not interesting to me when it carries a slogan underneath it."[25]

Additionally, there is the problem for the artist as activist regarding how to make work that engages with broad audience without resorting to populist cliché. As Baker points out, "the fact that [a good deal of contemporary animal art] can be so 'difficult to read' only exacerbates the problem of how effectively some of the artists who make it might address a subject such as the killing of animals."[26] It is thus notable that the work of Sue Coe was not included in any of these exhibitions and it is a matter for speculation whether the exclusion of Coe's work is due in part to the unfashionably direct and graphic nature of her work, which

---

[24]   Anderson, "Walk on the Wild Side," 466.

[25]   Glass, *Lisa Roet*, 10.

[26]   Baker, "'You Kill Things,'" 72.

has been described as "somewhat facile and overly literal."[27] The assumption here seems to be that art that is successful in communicating a message is often not so successful in more formal or poetic terms, presumably because form is made subservient to content.

Jane Scott questions whether the current interest by artists in the representation of animals could "indicate a general desire for a more literal approach in image making and the need to engage with a wider audience."[28] This suggestion is particularly pertinent to my current work and thinking. However, the use of the animal to signify something else in much of the artwork in recent exhibitions to which I presume she is referring does not seem to support this suggestion. Nor is it is clear why the depiction of animals should, in essence, make an artwork more accessible or attract a wider audience. Indeed, there is a worrying possible subtext for such a claim which is related to the trivialisation of animals as serious subject matter.

Tellingly, in his review of the Voiceless exhibition in the *Weekend Australian*, Sebastian Smee spent the majority of the review discussing the issues surrounding meat eating and factory farming, responding more to the catalogue essay by Ondine Sherman and opening address by J. M. Coetzee than to the artworks.[29] He "outed" himself as a meat eater, going on to state that "As an art critic, I don't see it as my job to pontificate in this column on the morality of meat eating …"[30] And yet, I would argue, it is this avoidance of a consideration of the ethics of our relationship with animals that will prove to be a weakness, not just in terms of art and exhibitions on the subject of human-animal relations, but in our society in general.

The generalised lack of consideration demonstrated by artists and curators for animals as sentient and self-interested individuals and the consequent failure to address the ethics of our relationship to animals is a key motivating factor behind my recent artwork, which seeks to aid in redressing this matter through the production of artworks that encourage the viewer to consider animals in a more empathetic manner and in doing so engage with the ethics surrounding human-animal relations.

Animals, Art, and Activism—A Personal Tale of Passionate Pursuits.

For almost as long as I can remember, art and animals have been twin passions that have ruled my life, but in the mid 1980s these passions collided when I began my career as an artist and, almost simultaneously, became actively involved in animal rights. Since that time these two pursuits have competed for my time and energy and my artwork and career as an artist have been affected both positively and negatively by my activist attitudes and activities.

---

    27   Adams, "Coldest Cut," 127. Adams does go on to state that "Now in the repoliticized early 90s, Coe really seems to be onto something."
    28   Scott, "Animal—Unsettled Boundaries," 4.
    29   Smee, "Beastly Goings On."
    30   Ibid.

My work as an animal rights activist involves trying to educate people about the suffering of animals at the hands of humans in the hope that they will either change their own habits or activities that contribute to this suffering and/or assist in lobbying for change. An essential aspect of this campaign work is the need to activate in people a sense of empathy and respect for animals. The position I hold regarding what I see as problematic about the relationship between most humans and animals is active, informed, considered, and passionately felt. It is a major part of my identity and has been of underlying importance to the subject matter of my artwork for almost 25 years. Despite this, for many years I avoided making artwork that overtly addressed this position. Eventually, around 2000, I began to confront the issues that I had been campaigning on as an activist for many years, with works such as the Model Animal and Dumb Animal series (Figures 9.1 and 9.2) using crude plastic models to explore the distanced, commodified relationship we have with "farm" animals. However, despite the fact that these works reflected my concerns about the treatment of animals at human hands, the issues were approached in an indirect manner, used a relatively "cool" aesthetic and played down the emotional aspects of the issues at hand. This approach reflected my concern at the time about making work that may be seen to be too didactic or polemical. However, in 2004, after explaining the concerns behind my artwork to a journalist, he asked me, why then, was I not making work that would try and help overcome these issues, rather than simply reflecting upon them? This important question prompted, or at least pre-empted, a significant shift in the manner in which my work deals with the subject of human-animal relationships, particularly with those animals that are used in food production.

The result has been that the key driving forces in my life—animals, art, and activism—have merged, resulting in artworks that address the core concerns behind my animal advocacy work more directly than ever before. The result of integrating these three elements is that, rather than creating artworks that are relatively passive in what they require of the viewer on an emotional level, I am now producing artworks that actively encourage the viewer to consider animals—particularly "farm" animals—as sentient beings rather than as insensate, objectified commodities. The intention has been to achieve this through the judicious use of anthropomorphism or, more aptly, "egomorphism" in the depiction of the animals. The term egomorphism was coined by Kay Milton as an alternative term for a common but inappropriate use of the word anthropomorphism. Milton argues that anthropomorphism can be misleading as it suggests that humanness is the departure point for any understanding of nonhuman animals, whereas egomorphism places the self, or ego as the primary point of reference. She also argues that anthropomorphism implies the attribution of human characteristics *to* other animals whereas as egormorphism allows for the perceiving of similar characteristics *in* animals.[31] The activation of egomorphism in the work thus reflects

---

[31] Milton, "Anthropomorphism or Egomorphism?" 255–71. Milton's term is superficially similar to Margaret King's term "ego-system" mentioned in Chapter 2. However, the two terms are actually opposed: King's refers to an anthropocentric and

Fig. 9.1   Yvette Watt, A Model Animal series, 2002, oil on linen, each work
              45 x 70cm. Courtesy of the artist.

the fact that this research is driven by a very personal empathy for nonhuman animals, and a consequent concern about human attitudes toward and treatment of other animals, especially those used for food, and this is reflected in the key role of the self in the works produced. The matter of my affinity with other animals has been expressed in a variety of ways in the artworks produced: by pairing self-portraits with animal portraits (Figure 9.2); by combining my features with those of other animals (Figures 9.3 and 9.4); through the use of my blood to produce images (Figure 9.5); and through the depiction of myself in conversation with "farm" animals (Figure 9.6).

Combining my features with those of animals is a gesture of association and solidarity and a metaphorical imagining of what their lives are like. By giving these animals specific human features—my features—and vice versa, giving myself animal features, I encourage the viewer to confront these hybrids as individuals and, even more importantly, as individuals that are hybrids of animals they might eat—as well as an animal (i.e., a human) that it is totally taboo to eat. Thus the

---

anthropomorphic view of nature, with the ego signifying very much the human ego; Milton's suggests rather an escape from this kind of view, with the ego indicating primarily an individual subjective response rather than a homogenised "human" one.

Fig. 9.2   Yvette Watt, Dumb Animal series, 2003, oil on canvas, each work
            180 x 270cm. Courtesy of the artist.

line between what can and cannot be eaten is blurred and the question of why it is acceptable to eat these animals is posed.

It is my intention that the artworks should deal with issues surrounding human-animal relationships in a way which is intelligible to a general audience, as well as being accepted as valid by a more specific contemporary arts audience. The engagement of anthropomorphism/egomorphism in my work has been instrumental in the achievement of this aim. Of key importance is the use of art as a tool for making socio-political comment and/or encouraging social change and, by extension, the role that images can play in affecting the way animals are thought about and hence treated. Baker addresses this issue in some detail in his 1993 book *Picturing the Beast*. As he points out,

> the stakes in representing animals can be very high. Who controls that
> representation and to what ends it will be used will be of profound importance
> in coming years as arguments over global climate change, disappearing
> and disfigured frogs, razed rainforests, hunting rights, fishing stocks and the
> precedence of human needs continues to build.[32]

---

[32]   Baker, *Picturing the Beast*, ix.

Fig. 9.3   Yvette Watt, three images from the Making Faces series, 2005, oil on
                2 panels, each work total size 27 x 36.5cm. Courtesy of the artist.

Consequently, I aim to present animals in such a way that viewers are caused to
question their own relationship to them, particularly those animals they probably
eat, in the hope of encouraging them to reconsider commonly held speciesist
attitudes toward animals. [33]

The recent artwork I have produced also responds to Steve Baker's important
question: "Can contemporary art productively address the killing of animals?"[34]
I would argue strongly for the affirmative and I would go so far as to suggest
that because the nature of art practice is more diverse now then it has ever been,
contemporary art can address the killing of animals more effectively than art from
previous eras. In addressing this issue, I have chosen not to use images of their
deaths. My rationale was to make artworks that cause the viewer to consider the
issues through having to *imagine* the lives and deaths of these animals, which
allows for more complex readings of the images. Ondine Sherman, in her essay
for the *Voiceless* catalogue, suggests the need for us to be confronted with pictures
of animals that show us the horrors of factory farms and the like. This is of course
precisely what animal activists such as myself do. In entering the places where
these animals are kept hidden, often enduring horrific conditions, our aim is to
document in photographs and film exactly how the animals are kept and to expose
their suffering. These images are for the most part raw, and untransformed, and
this is their strength. But at times they can be so shocking as to cause viewers to
turn away and refuse to engage with the image and hence the issue. I have chosen
therefore to avoid images of animal death that might shock or repel the viewers.
Although my artworks are issues-based and thus located within the scope of art
as a tool for socio-political commentary, I do not employ an overtly polemical
stance; instead, I attempt to lead viewers to question their own attitudes toward
animals in part by the use of humour in the work. However, the humour I have used
operates in a way that does not trivialise the animals. Rather, it is an unsettling,

---

[33]   The term "speciesism" was coined by British psychologist Richard Ryder as an
analogy to racism, sexism, etc., and denotes the privileging of humans above all other
animals, allowing us to treat them in ways we would not treat a human.

[34]   Baker, "'You Kill Things,'" 70.

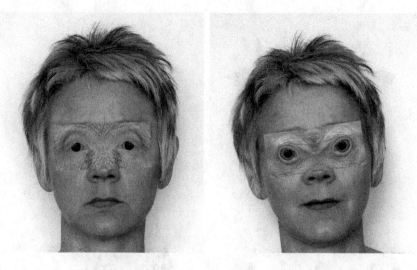

Fig. 9.4   Yvette Watt, two images from the Second Sight series, 2007/8, giclée print on photo rag paper, 65 x 58cm. Courtesy of the artist.

dark humour, particularly in those works that rely on a troubling pollution of the pure human form with animal features, such as the Second Sight series (Figure 9.4). In particular, I make judicious use of anthropomorphism/egomorphism and subtle humour to encourage an empathetic and hence compassionate response, thus prompting an acknowledgement of animals as active, self-interested agents rather than simply passive receptors of human ideas and actions. The notion of the animal as an active agent is particularly relevant in works such as the Domestic Animals series (Figure 9.6) which rely on the anthropomorphic idea of talking animals to engage the viewers with the issues at hand. But despite an apparent humorousness in these works, their subject matter is serious, a matter that is addressed by Erica Fudge, who notes that "a speaking animal can upset all kinds of assumptions by saying something we don't want to hear. Anthropomorphism can have an ethical dimension."[35]

The research is based on a firm belief in the role images can play in stimulating the imagination of the viewers in order to prompt them to consider the plight of the animals I depict in my work. This matter is addressed by novelist and former hunter Radclyffe Hall, who once said "I could no longer kill for the sake of pleasure ... I could in fact no longer ignore the victim, for imagination had led to understanding, and understanding to compassion."[36]

---

[35]   Fudge, *Animal*, 89.
[36]   Hall quoted in Kean, *Animal Rights*, 182–3.

Fig. 9.5   Yvette Watt, two images from Offerings series, 2007, artist's blood
              on linen tea towels, 49 x 71cm. Courtesy of the artist.

Implicit within Baker's question is another: can contemporary art enact social change? While I do not suggest that contemporary art can cause change by itself, I do believe that art can be a powerful instrument in the toolkit of social change. Like the multi-layered and targeted strategies of advertising—and animal rights campaigns—the production of artwork that directly addresses socio-political issues is one part of a much bigger and more complex picture. Contemporary artists have a reputation for questioning rather than blindly accepting the mainstream views of society. While this questioning drives the work of artists who engage with socio-political issues, it is only when the artist is prepared to make his or her personal position clear that the roles of artist and activist collapse into one. This merging of art and activism can provide for a productive forum for debate and discussion of the issues addressed by the artwork, both on an individual level between artist/ artwork and gallery audience, and in such fora as art journals and the general media. As someone who is both artist and activist, I am pleased to say that while I have encountered certain challenges, the combining of these two roles has proved to be effective as a campaign tool *and* beneficial to my art practice.

Ultimately, the questioning of human-animal relationships proposed by my work has far-reaching and profound consequences that extend well beyond simply treating animals more humanely. No doubt the implications of this questioning are behind the avoidance by so many artists (and the general public) of the ethico-

Fig. 9.6   Yvette Watt, two images from Domestic Animals series (top: *Culinary Tips*; bottom: *Art Lesson*), 2007, giclée print and ink on photo rag paper, 80 x 130cm. Courtesy of the artist.

political issues surrounding human-animal relationships. Nonetheless, my work is based on a conviction that the established attitudes toward animals must be changed and that art can play an important role in activating the shift in attitude that needs to happen for such change to take place. As W. J. T. Mitchell has pointed out: "The question of animal rights produces a combination of resistance and anxiety because to claim rights for animals entails a revolution so profound it would shake the foundations of human society."[37] It is this revolution that my works speak of, and I challenge those artists and curators who are truly interested in human-animal relationships to consider joining me in giving the foundations an enthusiastic shake.

---

[37]   Mitchell, "Rights of Things," x.

PART 3
Agency

Chapter 10
The Speech of Dumb Beasts

Helen Tiffin

Speechless

In the late eighteenth century, primal encounters between animals who had never experienced major predation and the most predatory of all species—humans—were not infrequent. In March of 1788, the *Supply* of Australia's first fleet made landfall at what their commandant named Lord Howe Island, a volcanic remnant which had over time become the home of numerous marine and terrestrial species of plants and animals, particularly birds. With arrivals by wind and sea from places as far distant as South Africa and New Caledonia, and survivals of the ancient continent of Gondwana (no longer surviving on the Australian mainland), Lord Howe Island had seen the evolution of more than 200 endemic species and sub-species, the most prominent of which in the terrestrial environment was the flightless woodhen (*Tricholimnas sylvestris*). Lord Howe's unique environment had achieved its own ecological stability, and when the noisy crew of the *Supply* came ashore—delighted not only to escape the confines of the ship but to hunt fresh meat after the restricted rations of the journey—woodhens came out in numbers to inspect the newcomers. Unacquainted with humans, they were tame and thus easily clubbed to death; but even this exertion on the part of the sailors proved unnecessary. By catching a few woodhens in their hands and breaking both their legs, the men discovered that the birds omitted a doleful cry which immediately drew others to them, thus facilitating their capture.[1] Such scenes, with different species, were played out wherever humans went, resulting not only in the slaughter of billions of animals, but in the extinction of thousands of species in an appallingly short space of time.

The encounter between the woodhens and humans is interesting in that, as members of the *Supply*'s crew recorded, it was a *particular* cry made by the suffering birds which brought the others. Yet at the same time, their torture and killing and the "sport" of the crew could be carelessly carried out since these were only animals—"dumb" brutes in both senses of the term—over which humans had, by predatory habit and Biblical fiat, been granted absolute rights. Yet the very separation of human and animal (the latter in such contexts having meaning only as the constitutive Other of the human) had always been (and continues to be) problematic. Even a century later, after Charles Darwin's

---

[1]   See for example, Hutton, *Lord Howe Island*.

Origin of Species challenged human exceptionalism on a number of counts, Max Muller could still claim:

> However much the frontiers of the animal Kingdom have been pushed forward, so that at one time the line of demarcation between animal and man seemed to depend on a mere fold in the brain, there is *one* barrier which no one has yet ventured to touch—the barrier of language ... [N]o process of natural selection will ever distil significant words out of the notes of birds or the cries of beasts.[2]

Yet such a confident pronouncement founders on precisely the relation it seeks to deny, since the very meanings of "language" and "beast" are not only historically and politically contingent, but in Muller's formulation necessarily stand for the basic difference *between* "human" and "animal"; the lack of what we term "language" necessarily signifying the obverse of the human.

Peoples we now consider "human" were once similarly "animalised" because of their apparent lack of complex consciousness, self-consciousness and, of course, language. Symptomatically, in a dramatic representation of an early encounter between European and Other, Shakespeare's Caliban does, in the eyes of Prospero, merely "gabble like a thing most brutish."[3] Prospero thus sets about teaching him language. Not, significantly, a *European* language but what Prospero understands as language itself. As Caliban accepts it, "you taught me language [not *your* language] and my profit on't is I know how to curse."[4] Native peoples encountered, like the fictional Caliban, by early European explorers were generally regarded as lacking language or possessing only very limited vocabularies largely if not exclusively concerned with the (animal) basics: sex and survival. The very instability (and co-dependence) of such categorisations both demonstrates and facilitates their contingent deployment in terms not just of species, but of race as well. As Cary Wolfe, citing Jacques Derrida (and Georges Bataille) puts it,

> [T]he humanist concept of subjectivity is inseparable from the discourse and institution of a speciesism, which relies on the tacit acceptance ... that the full transcendence of the "human" requires the sacrifice of the "animal" and the animalistic, which in turn makes possible a symbolic economy in which we can engage in a "non-criminal putting to death" (as Derrida phrases it) not only of animals, but of humans as well, by marking them as animal.[5]

"Animal" is thus an unstable category, entirely dependent for its meaning on what is seen to constitute "the human" in time and place, while "language" is that possessed by either "master" or "human"; rarely slave, servant, or animal. Just as Prospero teaches Caliban "language," so Robinson Crusoe in a later canonical text

---

2   Muller, "Science of Language," 271.
3   Shakespeare, *The Tempest*, 120.
4   Ibid., 121.
5   Wolfe, "Old Orders for New," 39.

of racial encounter does not, at least until the very end of the novel, acknowledge Friday's possession of his own language. Like Caliban, Friday learns the language of his master in order to serve him well. But even the acknowledgement of language possession does not guarantee that a subject is heard. As Gayatri Spivak has pointed out in relation to subaltern human groups, speakers may never be able to assume positions in which their speech can be heard as intelligible communication, let alone acknowledged.[6] Subaltern *positioning* itself can confer voicelessness on an erstwhile "speaker."

Representation and the Speaking Animal

Notwithstanding scientific and philosophical denials of the possibilities of animal speech, most human societies are familiar with speaking animals in a variety of written and oral genres, but most of these, whether or not they are anthropomorphic, have certainly been *anthropocentric*, and such representations of animal speech have generally been generically specific and thus circumscribed. Science, as the dominant paradigm of our times and authoritative arbiter of knowledge about animals,[7] represents animal behaviour, motivation, and communication within particular disciplinary parameters. Behavioural science may "translate" gestures and sounds for us, but never represents animals as speaking subjects. To even attempt this would so violate the protocols of the discipline as to disqualify the proponent as a scientist at all. Yet even as scientists themselves discover more about the complexities of animal interaction, inter- and intra-species communication, and determinative individual (and species) acts, animals are rarely or never credited with possessing "language." Like the so-called "sex and yams" theory of the speech of primitive peoples, animal communication is understood as confined to survival. Always linked to ideas of *human* cognitive complexity, experiments with speaking animals have been more or less confined to Primates[8] and conducted through the (human) *lingua franca* of Sign. Even so, for those scientists who credit, for instance, the findings of experiments such as those conducted on the iconic Washoe, there are numerous others who dismiss and decry its results or protocols. It is also worth noting, as the fictional Elizabeth Costello in J. M. Coetzee's *The Lives of Animals* does, that, like Caliban, Primates are expected to learn *our* language; there is little attempt on our part to learn theirs. Speaking of Wolfgang Köhler's experiments on apes, whose findings were published as *The Mentality of Apes* (1917), Costello speculates—on the basis of the way in which the exercises

---

[6]   Spivak, "Can the Subaltern Speak?" 271–313.

[7]   Latour likens the general Western public to the prisoners in Plato's Cave, with scientists going out into the light and coming back to the cave to interpret that world for us. Latour, *Politics of Nature*, 16.

[8]   Speech experiments have also been conducted on other animal groups. Attempts to demonstrate cognition through speech (rather than just a mimetic faculty) have been conducted on, for instance, the African Grey Parrot.

were conducted—on what the ape (Sultan) might have been thinking. Köhler's is not a *language* experiment as such; he is interested in mind, and the tasks he sets Sultan are aimed at a creature with presupposed limitations so that, as Costello notes,

> "At every turn Sultan is driven to think the less interesting thought. From the purity of speculation (Why do men behave like this?) he is relentlessly propelled toward lower, practical, instrumental reason (How does one use this to get that?) and thus toward acceptance of himself as primarily an organism with an appetite that needs to be satisfied."[9]

But if scientific representations of animals (and animal mentalities) generally challenge (or ignore) the potentials of animal speech, speaking animals are *de rigeur*, indeed pandemic, in fiction and oral tales for children. In children's literature it is almost axiomatic that animals speak on a par with humans, and that speaking animals and humans interact within the one story (for instance, in the Australian classic *The Magic Pudding* or the English *Pooh* series) or lead lives independent of human communities where inter-species speech is naturalised. Penguins, wombats, tigers, bears, frogs, and meerkats—plausibly or implausibly mixed—speak to each other freely. In *Little Red Riding Hood* the wolf has no problem imitating the voice of a (human) grandmother, or for a different wolf, communicating his nefarious intentions via language to the three little pigs. Children readily accept animal speech because, whether they are themselves kind (or cruel) to animals, they generally regard them—until they are taught otherwise—as equal companions and co-actants in the world.

But in imaginative literature for adults, animals are rarely represented as speaking; visual media—for instance, natural history programmes—may record animal sounds and increasingly hazard anthropomorphic interpretations, but this is not "speech." Speaking animals, it seems, vanish with childhood. Nevertheless, certain specific representational genres do employ—in fact depend, for their comic or satirical impact on—talking animals. In political cartoons,[10] fables, extended satire, and allegory, speaking animals again come to the fore. But in these genres they are necessarily human substitutes, deliberately anthropomorphised to present a (human) moral or illustrate human foibles. Satirists who attempt to both condemn human foibles *and* bring animals into presence in their own right often find that genre expectations, our ingrained habits of reading, almost always favour the human satirical message over, for instance, a critique of human treatment of animals. The case of George Orwell's *Animal Farm* is instructive here. As Jeffrey Moussaif Masson has pointed out, Orwell intended his work to be read not just as a satire on Stalinist Russia (and/or human democratic potential generally) but also as a protest against the treatment of farm animals.[11] Until recently, however,

---

[9]   Coetzee, *Lives of Animals* (Princeton UP), 29.

[10]   For more details see Baker, *Picturing the Beast*, 4.

[11]   Masson, *Pig Who Sang to Moon*, 9.

Orwell's work has almost always been read as a purely human satire, one in which we characteristically *read through* the animals to the humans for which they are merely masks.

Animals then, in Carol Adams's resonant phrase, become "absent referents"[12] not only of the meat on our plates, but through fable, allegory, comedy, and satire— genres which also depend for their effect on self-perpetuating stereotypes: greedy pigs, stubborn donkeys, silly cows. These are reductive portraits which depend less on (human-observed) animal behaviour than on prior textual representations. Categorised as "animals" (that is, the not-human) because they lack speech, and thus (we believe) associated complexity of thought, speaking animals in these genres, far from rectifying this perception, only reinforce it as they become doubly absent: their very capacity for speech ensuring their being read as transparent figurations of the human.

Genre Busters

As Nick Fiddes[13] and Steve Baker have both noted, our treatment of and attitudes to animals are wildly contradictory, rigorously compartmentalised to enable us to cope with our own attitudinal inconsistencies; and such compartmentalisation is reflected in representational genres. We divide "animals" into pets and pests, wild and tame, edible and inedible, and so on. In scientific writing, as has already been noted, animals do not speak and any signs of anthropomorphic portrayal are generally outlawed. In imaginative writing animals *can* be represented as having speech capacity, but outside of children's literature such representation is restricted to particular genres wherein we read through the animal mask to the "real" human message.

Recently, however, a number of writers and film and media producers have begun to deliberately breach such generic constraints. The film *Babe* offers a case in point. *Babe* succeeded in representing speaking animals (for both adults and children) in such a way as to not only bring animals into presence, but to decompartmentalise stereotypes and thus to destabilise one of our most fundamental strategies of self-deception. The "Three Little Pigs" who outwit the wolf may be cute, and small pigs are always popular in children's petting zoos. Yet most adults who can appreciate such "cuteness" in both representation and the "real" have no objection to eating "sucking pig" or ham. *Babe* presented audiences with a generic melange; crossing cartoon with realism, humour with tragedy, pigs with sheepdogs, pests with pets, edible with inedible, and human with animal, forcing us to (re)consider our habitual (and self-protective) categorisations and our inconsistent attitudes to and treatment of nonhuman animals. The transgressions

---

12 Adams, *Sexual Politics of Meat*, 14.
13 Fiddes, *Meat, A Natural Symbol*.

enacted in *Babe* resulted—at least for a short time—in a spectacular decline in pork consumption, especially in the United States.

For fiction writers whose traditional precursorial conventions are perhaps even stronger, such transgressions as those enacted in *Babe* can involve directly addressing this generic history in order to violate it. Not addressing such conventions either directly or indirectly can risk the work's dismissal as children's fiction or satirical and/or allegorical fable. Contemporary writers who wish to bring animals *as animals* into presence have thus employed a number of strategies to directly address (or circumvent) these generic expectations on the part of readers.

Yann Martel's *Life of Pi* does not include animal speech as such, but it requires its readers to believe in the possibility of a man (Pi) and a tiger (Richard Parker) spending (as a consequence of a shipwreck) more than 200 days together in a lifeboat in the Pacific Ocean. Tales of shipwreck almost inevitably invoke the stereotypes of the genre: thirst, privation, derangement, the drawing of lots, cannibalism, and rescue. *Life of Pi* has a double narrative framework. The bulk of the novel, as told by Pi himself to the unnamed narrator (or author, Martel), is concerned with Pi's childhood as the son of Pondicherry's zoo director, the family's decision to migrate to North America (with those animals now purchased by North American zoos), and the sinking of the freighter *Tsimitsum* not long after the voyage has commenced. The rest of Pi's tale (with the exception of a brief episode on a carnivorous algal island) consists of an account of Pi's voyage with Richard Parker, each survivor keeping to his separate territory in the lifeboat in a wary (but co-dependent) stand-off. After Pi and Richard Parker eventually drift ashore in Mexico, Richard Parker unceremoniously disappears into the jungle, while Pi recovers in Canada.

On behalf of the Japanese insurers of the *Tsimitsum* Mr Chiba and Mr Okamoto are sent to interview Pi about the sinking of the freighter. The Japanese agents refuse to accept Pi's account of survival in a lifeboat with an adult tiger. They want, as Pi diagnoses, "a story without animals."[14] Pi thus proceeds to summarily allegorise his own narrative wherein the animals of his main account become human equivalents, and the story metamorphoses into a stereotypical shipwreck narrative. Although his interlocutors "liked" the story with Richard Parker better, they did not "believe" it, finding a tale of human murder, cannibalism, and revenge much more credible, since it is what they expected to hear; a tale not of the restrained animality of Pi and Richard Parker, but of unrestrained human "beastliness." In formally structuring his novel to expose the strictures imposed on human/animal narratives by stereotype and reader expectation, Martel demonstrates the ways in which animal presence is occluded in narrative, in that only the human version of the tale—however bizarre and grizzly—is accepted and apprehended by listeners/ readers whose narrative expectations have been already conditioned.

Rather than addressing a particular narrative convention through which animals in fiction for adults are effectively silenced, Timothy Findley in *Not Wanted on*

---

[14]   Martel, *Life of Pi*, 324.

the Voyage and native Canadian author Thomas King in *Green Grass, Running Water* mix temporalities to address the issue of an originary loss of communication between humans and nonhuman animals. For Findley this is the textual instantiation of man's control over other animals in the Biblical *Genesis* (particularly the story of Noah and the Ark). For King, who also draws on the Bible and the Ark fable, loss of animal speech occurs via the Western invasion and settlement of North America, with its displacement and near-annihilation of native Canadian cultures, animals, their environments, *and* their stories. Recounting the story of Noah's Ark from the point of view of Mrs Noah (Noyes, in Findley's version) and the animals, Findley exposes the ostensible rescue of pairs of animals by Noah as, rather than an account of "salvation," one of violence, murder, and the destruction of the bond between humans and animals including that of shared speech. Once the Ark arrives on dry land, the rule of what Homi Bhabha (in another context) refers to as "Man Christianity"[15] is instituted as "law" for all time to come.

In the world before the flood, in the wonderfully biodiverse community Findley (re)conjures, species and racial divides do not exist. All species have their different ways, their different forms of interaction, but there are no fixed hierarchies. Moreover, strict dividing lines are deliberately problematised by the presence of in-between creatures: half-ape, half-human children; angels. The human Mrs Noyes can communicate with other animals, especially her cat Mottyl, and with the sheep she has taught to sing. This is not, however, a completely peaceable kingdom. Some individuals and species do not like each other, but they co-exist—if sometimes in states of "armed" truce. There are then, as in Tzvetan Todorov's formulation, differences in nature, but more or less equivalence in value.[16] In apparently acting on Yahweh's orders, Noyes/Noah destroys this world together with living entities we no longer regard as "real," such as fairies, demons, and unicorns, who are banished to the realms of the mythological after the flood. But it is the boundary-crossers, the so-called "Lotte children" (at least one of whom is Noyes's own child), who must be done away with if a definitive species boundary is to be effectively established and maintained. Since the very definition of humanity is the "not-animal," such half-human, half-ape hybrids cannot be allowed to exist. Findley's rewriting of the story of the Flood exposes the moment of instantiation of the species boundary, thus "denaturalizing" the apparently "given" division between humans and animals. Additionally, Findley imagines a richness of community outside the discursive constraints that we can observe being put in place in the course of Noah's voyage to reify humanity through the literal and figurative sacrifice of animals and the animalistic. And it is representation which is Noyes's chief weapon, just as much as the story of Noah's Ark in Genesis has itself operated in Western cultures to bolster patriarchy as well as to establish and maintain a strict human-animal divide.

---

15   Bhabha, "Foreword," xix.
16   Todorov, *Conquest of America.*

A flood is also the key motif in Thomas King's *Green Grass, Running Water*; and water provides the basis of both the original native Canadian creation myth *and* the narrative of a dispute over the building of a dam on native land in the twentieth century, the two stories interwoven in the text. In King's novel, present-day native Canadians meet Coyote the trickster God and re-animated fictional figures drawn from canonical European texts who appear in the novel as four Indians escaped from a lunatic asylum. All four—Robinson Crusoe, Ishmael, Hawkeye, and the Lone Ranger—are named after (or are) characters from texts where they are masters of another race, having relationships which are similar to that of the paradigmatic Crusoe and Friday.

Like Findley, King also re-writes the Biblical Genesis by shifting its base from the Middle East to North America where Coyote, a humorous and volatile figure analogous in position (but not character) to the Middle Eastern Yahweh of Findley's novel, is the bringer of both creativity and chaos. Here, rather than hosting Yahweh, Noah encounters Coyote, who, together with Changing Woman (the native American earth mother figure), proves a more formidable adversary than did Lucy, Ham, the animals, and Mrs Noyes in Findley's novel. "Man Christianity" is again challenged in King's novel, contrasting as it does with native beliefs *incorporating* humans (both women and men) as part of animal being and the wilderness rather than in opposition to them. As the "I" in the text explains to "Coyote," he is definitely not wanted on the Western/Christian voyage:

> "Wait, wait," says Coyote. "When's my turn?" [i.e., to speak]
> "Coyotes don't get a turn," I says.
> "In a democracy, everyone gets a turn," says Coyote.
> "Nonsense," I says. "In a democracy, only people who can afford it get a turn."
> "How about half a turn?" says Coyote.
> "Sit down," I says. "We got to tell this story again."
> "How about a quarter turn?" says Coyote.[17]

Pitting his version of religio-philosophy against the Biblical story, King replaces Eve and Adam with First Woman and Ah-dam, and the Prime Mover is instead the trickster/creator Coyote; at least until Coyote's (bad) dream gets loose and (the Old Testament and Christian) God also appears (by mistake). Arguing that it, and not Coyote, is "in charge of the world," the "silly Dream" wishes to *be* Coyote. Told this is impossible, but that it can be a dog, the Dream is not satisfied and gets "everything backward":[18]

> "That looks like trouble to me," I says.
> "Hmmm," says Coyote. "You could be right"
> "That doesn't look like a dog at all," I tell Coyote ...
> I am god, says that Dog Dream.

---

[17]   King, *Green Grass, Running Water*, 273.
[18]   Ibid., 1–2.

"Isn't that cute," says Coyote. "That Dog Dream is a contrary. That Dog Dream
    has everything backward."
But why am I a little god? shouts that god ... I want to be a big god!
"What a noise," says Coyote. "This dog has no manners."[19]

What E. M. Forster, in *A Passage to India*, refers to as "poor little talkative
Christianity"[20] is thus unleashed on native North America, muscling its way in
and claiming everything. In King's novel, animals not only speak; they resist
Ahdamn's naming:

> You are a microwave oven, Ahdamn tells the Elk.
> Nope, says that Elk. Try again.
> You are a garage sale, Ahdamn tells the Bear.
> We got to get you some glasses, says the Bear.
> You are a telephone book, Ahdamn tells the Cedar Tree.
> You're getting closer, says the Cedar Tree.[21]

(Tragically, however, Ahdamn is perhaps not entirely wrong: the conversion of
animals and trees into [manufactured] profit is the hallmark of the new [Western]
regime. The Bear will perhaps reappear, stuffed, at a garage sale, while the cedar
may indeed become paper.)

King uses comedy and irony in a juxtaposition of past and present to both
lament the passing of Coyote's world *and* to thwart readerly temptation to consider
a pre-Western ontology—in which animals were considered speakers in their own
right—as simple, childish, or primitive. Combining times past and present, fantasy,
and grim reality, King also succeeds in historicising Western/Christian attitudes
to animals and environment, thus relativising (and destabilising) naturalised
Canadian (settler) ontology and epistemology. "This is a Christian world, you
know," comments the narrative voice, "We only kill things that are useful or things
we don't like." In this world "[a]nimals don't talk. We got rules."[22] Deliberately
violating these "rules," King, in *Green Grass, Running Water* restores the potential
efficacy of pre-invasion Canadian cultures including animal agency and speech.

Barbara Gowdy's *The White Bone* is the most comprehensive (and perhaps
most controversial) of recent experiments in the use of anthropomorphism and
animal speech in a novel for adults. While Martel indicates the ways in which
animals "disappear" through their roles as human surrogates in fable, allegory, and
narrative stereotype, Findley and King reassess one of the most influential of Western
texts, the Biblical Genesis as a tale of (animal and environmental) destruction: the
instantiation of white (male) dominance over women and nonhuman animals, with

---

[19] Ibid., 2.
[20] Forster, *Passage to India*, 148.
[21] King, *Green Grass, Running Water*, 33.
[22] Ibid., 163, 123.

their effective silencing as subalterns. Barbara Gowdy also employs Christian and quasi-Christian symbology in *The White Bone*, but to a very different purpose.

While children's stories unashamedly espouse anthropomorphic animal portraiture, science dismisses anthropomorphic interpretations of animal behaviour and communication for, at times, quite contradictory reasons. In order to maintain what Bernard Rollin has termed the "moral agnosticism"[23] of an allegedly value-free science, science has generally denied similarities between animals and humans (except anatomically and genetically where such striking similarities can be more "safely" acknowledged). Contradictorily, other branches of science argue against anthropomorphism as denying animals a separate integrity. But as the 1987 American Veterinary Medical Association report belatedly acknowledges, "all animal research which is used to model human beings is based on the tacit assumption of anthropomorphism; and if we can in principle extrapolate from animals to humans, why not the reverse as well?"[24]

For Gowdy in *The White Bone*, as for Coetzee's fictional novelist Elizabeth Costello, the complex interplays of similarities and differences that inform our attitudes to animals can best be harnessed by the literary imagination: practised as writers are in extrapolating otherness from self—be it in terms of race, gender, ethnicity—the leap required to read observed animal behaviour is perhaps not as profound as we conveniently believe. And since our emphasis on differentiating ourselves from animals rarely stresses the anatomical or physiological, we prefer to designate animals as "lesser" through mentality, singling out those traits we regard as peculiar to ourselves—a practice which is anthropocentric and has often been ethnocentric as well. Such confidently drawn and self-fulfilling distinctions are also related, at least in the West (and Western-influenced) societies, to Cartesian dualism—the separation of mind and body—and it is these dualisms and the self-distancing ideologies which drive them, that are challenged by writers like Gowdy and Coetzee. In *The White Bone* readers are encouraged to empathise with the elephants through their very recognisable individual human-like characters, both good and flawed. Gowdy's elephants have mental and emotional complexity, particular forms of social solidarity, and individual and collective speech traits.

In a novel exclusively concerned with elephants like *The White Bone*, the representation of language is necessarily problematic. Jane Goodall or Birute Galdikas,[25] for instance, through their extreme familiarity with chimpanzees or orang-utans, can "translate" their sounds, looks, and gestures into a language we understand; but the animals themselves cannot be represented as communicating in English. In children's books concerned with animals or in adult satires, speaking animals are expected. But in a novel like *The White Bone*, representation of elephant language is both necessary in humanising the animals, yet dangerous in inviting infantilization or ridicule. Still, without a voice, without some direct speech, the

---

23   Rollin, "Scientific Ideology," 112.
24   Ibid., 113.
25   Galdikas, "Living with Orangutans"; Goodall, *In Shadow of Man*.

readers' inhabitation of the elephants' world would be strictly limited. *The White Bone* contains much direct speech, but interweaves dialogue with a third-person narration which can incorporate comments on that communication, reminding us that this is a form of translation from a very different vocal source:

> "I must tell you something," Torrent rumbled. "Several things. Vital … vital things … There are links you know nothing of."
> "Which links are these?" Tall Time said, affronted. Unconsciously he had dropped the formal timbre.
> Torrent jerked his head towards the She-S's. Trunk up, he took a long inhalation.
> "Any number of them," he rumbled …
> Torrent turned back around.[26]

While the elephants always use an English modified by their own particular point of view ("hindleggers" for humans; "roar-fly" for helicopter; "big grass" for bamboo; and so on), in this passage, where there are no such specific noun changes, Torrent does not "say," he "rumbles." Though the distinction is much less formalised in English than in some other languages, different kinds of address—formal and informal—are familiar to us. But here the words are not inflected to represent formal dialogue; it is the timbre that conveys the tonal-social distinction. Even in the phrase "Torrent turned back around" (instead of the more usual "Torrent turned around" or "back"), we have conveyed to us the sense of a very large body manoeuvring to speak.

Gowdy also makes speech only one of many forms of communication, others among which are of greater significance. Much of the novel is couched in the form of reportage of what is going on in the minds of Mud, Date Bed, and Tall Time. Some forms of elephant communication, such as infrasonic rumbling, group and individual rituals of mourning the dead, or touching each other with their trunks, have the imprimatur of scientific authority. Gowdy also attributes to her characters telepathic skills that we, as humans, lack. Mindtalking and visionary capacities are not shared by the entire group, but if the group's mindtalker is killed or lost, the gift passes to another member in a manner similar to that recorded in other animal groups when an individual with a singular role (or sex) is lost and another takes that place.

While the conversational exchanges of the elephants are rendered credible in this way, their speech allows human readers to identify more easily with them and to see them as individuals rather than a species, one of the most significant ways in which Westerners in particular distinguish between humans (individuals worthy of moral concern) and animals (a collectivity or, at best, species to be preserved or eradicated at human will). The anthropomorphism Gowdy employs in *The White Bone* is erected on the basis of natural history and/or scientific observations of elephants, and the acknowledgement of this basis—one made clear in the apparatus of the book as a whole—destabilises a reliance on singular genres. Moreover,

---

[26] Gowdy, *The White Bone*, 63.

she exploits the possibilities offered by fiction to address the question of other languages, particularly in relation to animals. After all, animals are never without language even if we prove unable to translate their speech.

In *The White Bone* the elephants, represented as having complex spiritual and telepathic connections with their surroundings, possess a consciousness that arises out of their extraordinary capacity for memory, one which well exceeds the human. The elephants think that this accounts for their size: "that under that thunderhead of flesh and those huge rolling bones they *are* memory." When their "memories begin to drain their bodies go into decline, as from a slow leakage of blood."[27] Even humans who know virtually nothing of elephants know that they "never forget." This is true, Gowdy notes in her prologue, because the slow process of memory draining is almost never allowed to occur because "nine out of ten are slaughtered in their prime long before their memories have started to drain. I speak of the majority, then, when I say it is true what you've heard; they never forget."[28] For elephants, however, memory, regarded by us as a function of mind, is not a capacity that can be separated from the body, while metaphysical speculation and dreaming are also products of the same embedded sensory experience:

> [E]very odour they have ever sucked into their trunks, every flicker of sunlight they have ever doused with their tremendous shadows is preserved inside as a perfect and instantly retrievable moment. They rarely ask, Do you remember? The remembering is taken for granted. It is the noticing they question: Did you smell that? Did you see it? ... The precise tenor of the wind that lowed in the acacias that day, how the sun slammed down through the foliage ... Suppose, off to one side, waves of salt dust had swirled up from the pan. In memory, they can turn their gaze on the waves and ponder this phenomenon of a lake bed dreaming its lost lake.[29]

While human memory also invokes senses other than sight, and leads to speculation upon the past, we associate this process with mind rather than with body. Gowdy's novel deliberately breaches the mind/body dichotomy that forms the basis of much of Western understanding (as well as its speciesist, racist, colonialist, and gender biases and prejudices). Even though we are sometimes forced to acknowledge the indissoluble embeddedness of mind in body, that reluctant acknowledgement is indicative of the persistence of this basic Western separation. In *The White Bone*, by contrast, the huge bulk of the elephant body, its physiological functions, the desires and sufferings of fleshly being, are represented as intrinsic to the complexities of mind. Although it was René Descartes who most famously encapsulated the Western division of mind and body in his *cogito ergo sum*, such a separation was already a part of the West's philosophy and religion in the works of Aristotle and in early Christian thought. To become fully human was to transcend one's animality, one's earthbound substance. *The White Bone* challenges this construct

---

[27]   Ibid., 1.

[28]   Ibid.

[29]   Ibid., 1–2.

which has been historically instrumental in claims to racial, gender, and species superiority and has underwritten the separation of (human) being from (extra-human) environments as well.

Elizabeth Costello in Coetzee's *The Lives of Animals* uses the occasion of her lecture at an American college to deny this separation and to draw attention to the ways in which animals and the animalistic have been regarded and treated by (Western) humans. Speaking of Descartes's claim that animals were mere automata, machines, Costello suggests the following:

> "Cogito ergo sum" he [Descartes] also famously said. It is a formula I have always been uncomfortable with. It implies that a living being that does not do what we call thinking is somehow second-class. To thinking, cognition, I oppose fullness, embodiedness, the sensation of being—not of consciousness of yourself as a kind of ghostly reasoning machine thinking thoughts, but on the contrary the sensation—a heavily affective sensation—of being a body with limbs that have extension in space, of being alive to the world. This fullness contrasts starkly with Descartes's key state, which has an empty feel to it: the feel of a pea rattling around in a shell.[30]

In her two lectures on the lives of animals, Costello also addresses the function of writing, paying particular attention to the kind of self-serving philosophical fictions promulgated by Descartes. Such philosophical fictions are different in each age, Costello suggests, but they nearly always emerge in practice as being to human benefit at animal expense.

In counteracting Cartesian dualism through the representation of animals *as animals*, Gowdy and Coetzee also work towards undermining the bases of racial and gender inferiority, confirming this inseparability of mind and body in both the material and metaphysical realms. Inextricably connected as it has been to concepts of animality, this triple lack—language, consciousness, and mind—is best interrogated by the kind of imaginative writing that questions, through the ways in which it represents animals, the dominant science paradigms whose contrasting generic approaches deliberately foreclose knowledge other than their own. Science has been instrumental, these two writers suggest, in giving us our current ideas of animality—ideas that classical philosophical theories and Christian doctrine reinforce.

In *The White Bone*, the elephants' search for the "that-way Bone" echoes religious quests in novels about humans; and together with the elephants' belief in the "She" and their reading of signs, these traits accord the animal a complex spiritual sense usually only attributed to humans. But by including this quest motif (and its quasi-Christian symbology), Gowdy risks invoking an allegorical reading of the text along Christian lines,[31] notwithstanding her strategy where humans,

---

[30]   Coetzee, *Lives of Animals* (Princeton UP), 33.

[31]   Date Bed is read as a "Christ Figure" by Neta Gordon in "Sign and Symbol in Barbara Gowdy's *The White Bone*."

destructive, violent, and disruptive as they are of elephant being, are also depicted in the text, though generally "offstage," erupting apparently out of nowhere to perpetrate a savage violence. Gowdy, however, further insures against our reading the elephants as human surrogates by having the elephants themselves sometimes question the efficacy and purpose of symbolic and allegorical readings, omens, and events; Gowdy's speaking elephants are elephants, not humans-in-disguise.

Speech in Strange Tongues

Imaginative writing, as Elizabeth Costello argues in *The Lives of Animals*, can go where (human reason-based) philosophy fears to tread. In short, it offers a mode in which the falsely unified and allegedly speechless category of the animal may find voice. Fiction writers have constantly to imagine radical otherness; to write dialogue for characters they can only imagine. Elizabeth Costello argues that this is best achieved in the case of animals by emphasising not mind, as separate entity, but the embodiedness of all being; mind itself being inevitably embodied. Nevertheless, constraints imposed by traditional literary genres and naturalised habits of reading the animal subject in imaginative works often necessitate a direct address to these generic constraints before animals may "speak" to adult audiences.

Contra Max Muller, many scientists are now "distill[ing] significant words out of the notes of birds and the cries of beasts." Yet notwithstanding such moves in science and fiction, fundamental philosophical problems do remain, as novelist Coetzee and philosopher Jacques Derrida have pointed out. We can only imagine the languages of animals[32] within the confines of our own; there are inherent difficulties in entering into any *dialogue* with our animal "Others," given that the Other is precisely what is both supposed and repressed by language itself. A "linguistic carnivorousness,"[33] to use Coetzee's term, thus effectively traps nonhuman Others with our own words, rendering them (independently) speechless.

In *Animal Rites: American Culture, The Discourse of Species and Posthumanist Theory*, Wolfe cites Ludwig Wittgenstein's observation that "If a lion could talk we could not understand him" as a particularly apposite "epigraph to the debates that have taken place over the past century on animals, language and subjectivity." Nevertheless, Wolfe argues, Wittgenstein's statement, "[H]as often been misunderstood—I'm not sure I understand it myself—and it is only complicated by Wittgenstein's contention elsewhere that to imagine a language is to imagine a form of life."[34] But as Wittgenstein continues,

---

[32]   The very category "animal" is absurd, usually denoting, broadly, the not-human as Derrida (amongst others) has noted.

[33]   See J. M. Coetzee, "Meat Country," 42–52.

[34]   Wolfe, *Animal Rites*, 44.

It is, however, important as regards this observation that one human being can be a complete enigma to another. We learn this when we come into a strange country with entirely strange traditions; and, what is more, even given a mastery of the country's language. We do not *understand* the people. (And not because of not knowing what they are saying to themselves.)[35]

Vicki Hearne, Wolfe notes, finds that "the lovely thing about Wittgenstein's lion is that Wittgenstein does not leap to say that his lion is languageless, only that he is not talking." The reticence of the lion, so Hearne observes, "is not the reticence of absence, absence of consciousness, say, or knowledge, but rather of tremendous presence" of "all consciousness that is beyond ours."[36]

We are surrounded by these consciousnesses beyond ours, beyond our limited understanding, and, with the exception of a handful of rare (human) individuals, have only recently acquired a more urgent need to attempt to understand; while at the same time being forced to acknowledge the limits that animals' potential bringing into presence necessarily exposes in ourselves. Even as animal consciousness is both "supposed and repressed" by *our* language(s), imaginative writers are increasingly addressing the task of giving us glimpses into these realms, redrawing these animal languages in fictions which not only challenge former and formal generic boundaries, but which reach to address the question of those vast possibilities beyond our own language and understanding.

---

[35] Wittgenstein quoted in Wolfe, *Animal Rites*, 44.
[36] Hearne quoted in Wolfe, *Animal Rites*, 45.

Chapter 11
Extinction, Representation, Agency: The Case of the Dodo

Carol Freeman

Images of dodos are everywhere on the island of Mauritius. In the capital, Port Louis, ceramic and plastic replicas crowd market stalls, rolls of fabric are adorned with borders of the extinct bird, cushion covers feature the familiar plump profile, fluffy dodo toys accentuate a generous form, while a local wood-carver paints soulful eyes on a pine version. On a quiet beach at Gris-Gris on the southern coast of the island, a weathered sign in the shape of the bird warns *Bain Dangereux* (dangerous bathing). However, the logo of Air Mauritius, one of the many airlines that fly tens of thousands of tourists from Europe to Mauritian resorts during the holiday season, is the elegant red-tailed tropical bird or *paille-en-queue* from Reunion Island to the north of Mauritius. As explanation, one of the airline's advertisements with the wry caption "For three hundred years, there's been no competition" and the promise of "Non Stop Caring" for passengers, has an outline of a stout dodo defined by words printed in the shape of the bird:

> The poor old dodo, Dronte or Didus heptus, big and ungainly, lumbering and unable to fly, was an unfortunate Mauritian native that became extinct before the start of the 18th century—unlike its fleet-feathered, high flying compatriot, the Paille-en-Queue, which is alive and well and winging its way merrily above the seas surrounding this jewel-like island. Little wonder then, that Air Mauritius chose the Paille-en-Queue as our symbol: it just goes to show that when it comes to flying, we're certainly not dodos.[1]

The creative director of Mauritian advertising agency Publico acknowledges that the dodo has a "bad" image and, while a representation of the species called Tizan was the Mauritian team's mascot at the Indian Island Games in 1985, he points out the dodo was not a contender for mascot when a team went to the Olympic Games in Athens in 2004.[2] But what if the species' image changed? This is now a possibility, as research by Andrew Kitchener, curator of birds and mammals at the National Museums of Scotland, has provided evidence that dodos were "lithe and active" birds that could run fast[3]—a perfect mascot for an Olympic team.

---

[1] Air Mauritius, "For three hundred years, there's been no competition," print advertisement, Cerca East Shearer, South Africa, c. 2000.

[2] François Lam, personal communication, 15 August 2003.

[3] Kitchener, "Justice at Last," 27; see also his more detailed version of this article: "On the External Appearance of the Dodo."

History and the Dodo

In popular perception the name "dodo" has long been synonymous with extinction, especially with extinction caused by human activity. The publication of Lewis Carroll's *Alice's Adventures in Wonderland* (1865) at the same time as a grave of dodo bones was unearthed on Mauritius assured that the species was inextricably linked with both humour and death. In the twentieth century, constant references reinforced these associations. For instance, the dodo features in the title of David Quammen's book *The Song of the Dodo: Island Biogeography in an Age of Extinction* (1996), and images of dodos are used by environmental organisations that promote the protection of endangered species, such as the Durrell Wildlife Conservation Trust and the Jersey Zoological Park. In countless phrases, logos, allusions, and naming, the figure of the bird has been linked with slowness, stupidity, and things that are defunct, amusing, or out of date—the earliest record of the phrase "as dead as a dodo" refers to May, 1881.[4] Ideas about the bird's appearance and demise were rarely questioned in early scientific and popular writing and are usually accepted today in general and even academic literature. Indeed, notions about the dodo's form have become entrenched, even mythological, and until attention is firmly focussed on the animal subject rather than human representations of the species, public perceptions are likely to be dominated by these long-established and often erroneous understandings.

The Animal Moment that is being felt in both academic and wider circles offers opportunities to re-examine and reassess animal representations. Studies in this new field have resulted in considerations of issues of animal agency and innovative intersections between academic disciplines and approaches to the subject. This is what happened on the H-Animal website in 2006 when animal studies scholar Erica Fudge critiqued historian David Gooding's paper "Of Dodos and Dutchmen: Reflections on the Nature of History." Gooding's argument in the paper rests to a considerable extent on a quote from an unnamed artist who described the birds as "allowing us to beat them to death."[5] Fudge concentrates on Gooding's statement that the story of the dodo was merely an event in "unhistorical time"; that is, a history of nature that is not "willed" as human history is. She opens up a debate about the place of the animal in history and narrative.[6]

However, Gooding's paper is problematic on a number of other levels. Firstly, his premises about history are based on a set of assumptions and ideas that have little relevance in the twenty-first century when scholarship and research methods have become broader and infinitely more complex. Gooding concentrates particularly on what historian R. G. Collingwood refers to as "scientific history" (1946)[7] and fails to take account of recent research in a number of disciplines

---

4　　Martin, "Phrase Finder."

5　　Gooding, "Of Dodos and Dutchmen," 39.

6　　Fudge, "History of Animals."

7　　Collingwood, *Idea of History*, 8–10.

that stresses how humans constantly interact with nonhuman animals and are enormously influential in their actions toward animals. This includes work on human-animal relations by Jennifer Wolch and Jody Emel, and Sarah Whatmore and Lorraine Thorn in the 1990s, not to mention the biologists and zoologists who have specifically investigated the history of the dodo and have suggested new perspectives on specific issues dealt with in his article.[8] Indeed, Gooding ignores major points Collingwood makes about history *per se*, including that attention should be drawn to the historian's position in time. Collingwood accepts the possibility of evolving disciplinary methods and viewpoints, admits that the very idea that history is about the "actions of human beings that have been done in the past" is controversial, and refers to historical thought that is "out of date" and that demonstrates the "illusion of finality."[9] Gooding's reading of Collingwood's work is selective and reproduces anthropocentric ideas while, in fact, many of Collingwood's observations implicitly encourage new approaches to history that are particularly pertinent to historical studies of animals today.

The aspect of Gooding's paper that I want to deal with in this chapter, however, is one that is fundamental to his argument and Fudge's critique. It concerns the representation of the dodo and the way the species' history has been written. Gooding discusses a version of the extinction story as if it were the only narrative that has been suggested. I will contest his account of the form, habits, and behaviour of the bird, as well as how and why the species became extinct, evaluating both historical evidence and current research on the subject that has taken place in both science and the humanities. This multidisciplinary approach is essential to understanding the slipperiness and complexities of the animal subject, especially in a historical context.

One of the problems with Gooding's paper is that the bases for his assumptions about the dodo are all from popular or concise sources. He seems to have gained knowledge through "ordinary" channels that Collingwood considers are superficial and "invariably out of date."[10] First Gooding, writing in 2005, sites Errol Fuller's popular book *Extinct Birds* (1987), ignoring a much more detailed work specifically on the dodo by the same author published in 2002; then he references a book called *A Guide to the Pigeons and Doves of the World* (2001), an encyclopaedic work designed for field identification with very brief, jumbled evidence about the bird's history; and finally two descriptive books on the history of Mauritius (one described as "concise") published in 1977 and 1998. Most zoologists would consider these references an inadequate framework on which to base an academic argument, and Gooding fails to properly address the interdisciplinary nature of the subject.

---

[8] See Wolch and Emel, *Animal Geographies*; Whatmore and Thorne, "Wild(er)ness"; Philo and Wilbert, *Animals Spaces*; and works by zoologists cited later in this essay.

[9] Collingwood, *Idea of History*, 7–10.

[10] Collingwood, *Idea of History*, 7.

The representation of the dodo that Gooding uses is paraphrased by Fudge thus: "the accepted narrative in which dodos lived peaceful lives on Mauritius, until the sixteenth-century arrival of human predators, and were decimated by these new arrivals against whom they had no form of protection."[11] My concern is that Gooding does not question this account and Fudge offers no specific alternative. Gooding uses it to support his suggestion that the discipline of history is concerned with a historical and unhistorical time: "human history is actually opposed to nature as if the non-human world was a passive background ... an active versus a passive principle."[12] He goes on to do the same himself, writing the dodo out of his version of history, rather than alluding to a more complex and less definitive story. By contrast, Fudge references very recent human-animal studies writers, such as Nigel Rothfels, Jonathan Burt, and Virginia DeJohn Anderson, who are now integrating animals into history narratives and according them agency—for instance, "a role in shaping human culture and landscapes."[13] But her essay is brief and does not pursue her implication that dodos may have been active agents during the process of extinction.

I argue that the first and most important requirement in any discussion about extinct species (and many living animals) is to rethink current understandings of them, rather than concentrate on methods of history-making and human intellectual concerns. The new wave of human-animal studies scholarship demands a thorough examination of existing assumptions and the analysis of a wide range of sources and recent discoveries and theories. As Ken Shapiro outlines in a recent human-animal studies policy paper:

> [V]iews of nonhuman animals in general and of a particular species or class of animal are often prejudiced and anthropocentric, consisting of layers of ideological and linguistic biases that serve only human interests. By unravelling and examining the layers of these social constructions, scholars reveal, as much as is possible, the animal as such—the animal seen and related to directly and without prejudice.[14]

In the case of the dodo, these layers include first-hand written reports, visual representations, and archival material. In addition to these, recent scientific assessments of physical remains and consideration of the politics of human-animal relations offer surprising insights. As with postcolonial readings and feminist histories, when the animal is placed at the centre of the study and the historical evidence is approached from different, relatively unbiased perspectives, new narratives result. This essay critically re-examines representations and reports of the dodo's extinction over four hundred years. It attempts to revalue dodo experience and redress imbalanced representations of human and nonhuman

---

11 Fudge, "History of Animals."
12 Gooding, "Of Dodos and Dutchmen," 41.
13 Fudge, "History of Animals."
14 Shapiro, *Human-Animal Studies*, 1.

animals in the light of old and new evidence. I ask: where and how was agency actually manifest in the case of the dodo's extinction?

Representations of the Dodo

The dodo (*Raphus cucullatus*) was endemic to Mauritius, a small island in the Indian Ocean originally uninhabited by humans. This island is one of three in the Mascarene group, first visited by the Portuguese in 1510, claimed by the Dutch in 1598, and later occupied by French and British colonists. *The Oxford English Dictionary* states that the word "dodo" originates from the Portuguese *duodo* or "simpleton," "fool." But this is by no means certain, and other possibilities include a Dutch derivation—*dodaersen* translated as "fat behinds," or *dodoor*, which means "sluggard," or *walghvogel*, which means "loathsome bird" or "nauseating fowl" in reference to the taste of the bird's flesh.[15] None of these meanings are complimentary, and many are ideologically biased. They demonstrate a power relation that privileges the human namer and consumer. There is also an unsubstantiated claim that "dodo" is an onomatopoeic version of the bird's call, a two-note sound like "doo-doo." But it is human language and visual images, rather than the dodo's own call, which dominate traditional ideas about the species.

Most sources agree that the dodo disappeared during the late 1600s, less than a hundred years after human settlement of Mauritius. Like many islands, it had a unique and vulnerable biota that included no mammals or predators. What evidence there is suggests that the species was common in areas close to the coast rather than inland, a fact that would have favoured their extinction by human visitors, but even this is uncertain. Palaeontologist Julian Hume comments "much of the information [about the dodo] has been derived from few genuine but inadequate contemporary accounts and illustrations, yet a wealth of assumptions and over zealous mis-interpretation about dodos' ecology and morphology has taken place." Biologist Ralph Dissanayake states "the image of the dodo in representation is often a caricature." Errol Fuller, natural history artist and writer, also maintains "what is actually known of the living breathing dodo is minimal," while Samuel Turvey, zoologist specialising in human-caused extinctions and Anthony Cheke, an ecological historian, warn against "simplistic teleological interpretations of history" in regard to the dodo.[16] In other words, common understanding of what a dodo looked like, and especially the species' habits and history, is open

---

[15] This and other basic information concerning the species' history is repeated in a number of sources: Hume, "History of the Dodo," is the most scholarly and detailed of a number of recent re-examinations that include Dissanayake, "What did the Dodo Look Like?"; Turney and Cheke, "Dead as a Dodo"; and popular but detailed histories by Fuller and Den Hengst.

[16] Hume, "History of the Dodo," 65; Dissanayake, "What Did the Dodo Look Like?" 165; Fuller, *Dodo*, 25; Turvey and Cheke, "Dead as a Dodo," 152. For more on the unreliability of early accounts of the dodo see Cheke, "Establishing Extinction Dates."

to question. This means that the validity of Gooding's premises about the dodo's extinction—what Fudge describes as "the accepted view"—stands on very shaky ground. Traditional ideas about many extinct or colonial animals are often based on scant or prejudiced information. First-hand accounts are often contradictory, others are recorded long after sighting, clearly fanciful, informed by cultural mores or assumptions, and frequently impossible to verify. Few live specimens of the dodo were sent to Europe or the East, there was no systematic investigation of the bird's habits, and little skeletal material remains. In the light of new discoveries and changing attitudes toward animals, then, this essay takes up Collingwood's idea of constant questioning and awareness of the historian's position in time, as well as changing methods and viewpoints, and re-examines the so-called "first-hand" accounts of the dodo that survive.

The first extant written descriptions of the bird appeared in 1599 and 1601 following visits to the island by Dutch expeditions led by Jacob van Neck and Heyndrick Jolinck. They were described as having "wings as large as of a pigeon, so that they could not fly and were named penguins by the Portuguese."[17] Another report maintains they were "twice as big as our swans," having three or four black (or yellow) quills instead of wings, and a few curled grey plumes in place of a tail."[18] However, an account of the only dodo definitely known to have reached England states that the species was stouter than a turkey, but had a more erect shape.[19] Other reports comment that dodos had long legs and, significantly, that "they walked upright on their feet as though they were a human being."[20] These descriptions are notable but rare; they appear only fleetingly and are seldom mentioned in secondary sources. An illustration of Dutch activities onshore, pictured in 1601 (Figure 11.1), includes the first known visual image of a dodo under the tree on the middle left. This image was probably produced in The Netherlands from the written accounts with which it corresponds, rather than originating from a drawing made on site. This bird is also relatively slim and has long legs; however, the descriptions include no measurements that would help establish the dodo's proportions.

The details of the size and edibility of dodos that are given are often vague or inconsistent and can be easily misinterpreted. One report says: "These particular birds have a stomach so large that it could provide two men with a tasty meal."[21] Other reports say that much of the dodo's flesh was tough, even when cooked for a long time, with only the stomach and breast edible, so doves and pigeons were preferred. Most reports imply that dodos were unpalatable. The idea that the species was routinely bludgeoned to death may have arisen from an indistinct engraving and an equally enigmatic poem published in 1648 that mentions "round-rumped

---

[17] Heyndrick Jolinck quoted in Hume, "History of the Dodo," 67.

[18] Wybrant Van Warwijck's report quoted in Fuller, *Dodo*, 51–2.

[19] Hamon l'Estrange quoted in Fuller, *Dodo*, 69.

[20] François Cauche quoted in Fuller, *Dodo*, 68; anonymous quoted in Fuller, *Dodo*, 53.

[21] Van Warwijck quoted in Hume, "History of the Dodo," 67.

Fig. 11.1   Anonymous engraving from *Het Tvveede Boek* (1601).

dodos they destroy / the parrot's life they spare that he may peep and howl"[22] and on several reports that briefly mention visiting sailors killing dodos. Two of these accounts use words so similar to earlier ones, such as an ambiguous statement in 1611 that "the Dutch have been catching and eating [dodos] daily and not only these birds, but many other kind ... which they beat with sticks and catch," that it is likely they were copied rather than the incidents observed. But the writer of the 1611 account, Pieter Verhoeven, also states that sailors must take care that dodos "do not bite them on the arm or leg with their great, thick, curved beaks."[23] This capacity for biting back is never mentioned in the version of human-dodo interactions that is generally accepted today, and yet its singular nature implies it may be based on the report of an animal "seen and related to directly and without prejudice,"[24] rather than copied from previous accounts.

In reports from 1606 and 1628 other sources of food on Mauritius are mentioned, such as hogs and cows, as well as a large numbers of rats, monkeys and goats—introduced animals that threatened the dodo's survival and hastened the species' extinction. And then, in 1662, shipwreck survivor Volkert Evertszen reports that the only remaining dodos, on a small island off the Mauritius coast, were fast runners and adds "one of us would chase them so that they ran towards

---

[22]   Willem Van West-Zanen quoted in Fuller, *Dodo*, 56.
[23]   Pieter Verhoeven quoted in Fuller, *Dodo*, 59.
[24]   Shapiro, *Human-Animal Studies*, 1.

the other party who then grabbed them. When we had one tightly gripped around the leg it would cry out and then the others would come to its aid and they could be caught as well."[25] This account is sometimes discredited because it was written seven years after the event and cites behaviour typical of another Mauritian bird, the red rail.[26] But Cheke refutes this idea, while Kitchener suggests a similarity between the habits of birds endemic to Mauritius, so Evertszen's account remains a tantalising image of a very different bird to the stereotypical large, fat, slow animal that is central to the "accepted narrative."[27] Ambiguity, contradiction, frustratingly incomplete accounts, and questionable observations are features of dodo reports, as they are in reports of many other animals now extinct. As early as 1848, H. E. Strickland and A. G. Melville, the writers of the first comprehensive study of the species, noted that "the facts recorded by ... witnesses have been transcribed and often confounded by a multitude of compilers."[28] Cheke points out that there was also a lack of specimens, particularly in France, that led many late eighteenth- and early nineteenth-century French scientists to doubt the very existence of the bird, and that there was generally little interest in the dodo's decline in comparison with other Mascarene species.[29] Factors such as these, then, make the writing of extinction histories a particularly difficult matter and highlight how the "facts" on which we base "knowledge" of an animal are often unreliable.

Visual images are just as enigmatic. Paintings of dodos appear relatively early and there are many of them. A work by court painter Jakob Hoefnagel in 1602 was previously accepted as modelled on a living bird from the menageries of Emperor Rudolph II at Prague, but Hume and Fuller believe it was made from a badly stuffed specimen.[30] They point to the sunken eyes, blackish skin and withered head as denoting partial decomposition, and twisted wing feathers indicating bad drying techniques. The bird's tail also appears lost and the legs may have been painted from another source. The use of specimens as models for zoological illustrations was a consistent practice prior to the late 1800s, when photographs of animals became common. This representation is an example of how taxidermists and artists, many of whom were untrained, contributed to the construction of ideas, and suggests that illustrations used as primary sources of scientific knowledge should be consistently interrogated.

The well-known painting by Rowlandt Savery dated 1626 is probably modelled on a white dodo that was also part of a large collection of unusual specimens, as well as live animals, in the menagerie in Prague (Figure 11.2).[31] This figure, which ironically has two left feet, is one of a series of dodo paintings by Savery

---

25 Quoted in Fuller, *Dodo*, 72.
26 Hume, "History of the Dodo," 69; Fuller, *Dodo*, 73.
27 Cheke, "Establishing Extinction," 156; Kitchener, "Justice at Last," 27.
28 Strickland and Melville, *Dodo and its Kindred*, 8.
29 Turvey and Cheke, "Dead as a Dodo," 151.
30 Hume, "History of the Dodo," 72; Fuller, *Dodo*, 81.
31 Hume, "History of the Dodo," 72.

Fig. 11.2   Engraving of a dodo for the *Penny Cyclopaedia*, after Rowlandt
Savery. In Strickland and Melville, *The Dodo and its Kindred*,
1848.

that influenced Tenniel's illustration in *Alice in Wonderland* and contributed so
overwhelmingly to what is generally believed to be the shape and stance of the
dodo, although Tenniel's dodo is not quite so stout and is relatively upright. Today,
Savery's image is used in advertising and appears on numerous souvenirs and
brochures in Mauritius, as well as circulating globally in books and on websites.
It has provided reinforcement for the common belief that dodos were plump,
clumsy, unwary, and easy to capture and it is this perception that is repeated in
David Gooding's paper, along with the idea that members of the species could not
defend themselves and were hunted and killed in large numbers. Indeed, Gooding
states "the numerous depictions of [dodos] by European artists have meant that
the appearance of this exotic bird is well known."[32] However, it is not the *shape* of
the bird, but Savery's painting that is so familiar. There are images of the dodo in
existence that are not well known and are rarely researched, discussed, or analysed
outside the discrete field of zoological illustration. They tell a different story.

---

[32]   Gooding, "Of Dodos and Dutchmen," 38.

The first drawings of live or recently killed birds on Mauritius were made by Joris Joostensz Laerle, a skilled professional artist, during a visit to the island in 1601 (Figure 11.3). These images are considered by Hume and Fuller to be among the most reliable representations of the dodo.[33] Note the long legs, upright stance, and relatively small body. Another image was made from living dodos in the menagerie of the Great Mogul Emperor in Surat, India, in 1625. These birds were described in *The Travels of Peter Munday* as "big-bodied as great Turkeyes, covered with Downe, having little hanguing wings like shortt sleeves, altogether unuseful to fly withall, or any way to help themselves."[34] On the other hand, an anonymous Dutch sailor on his way to Surat in 1631 described them as "superb and proud, with stiff and stern faces and wide-open mouths." He felt that dodos had a "jaunty and audacious gait" and that "they would scarcely move a foot before us."[35] This alternative perception of dodos implies that their behaviour was lively, brave, or bold and that they were resistant, rather than stupid. The sailor's report is framed by a respectful attitude and a fresh and observant approach, rather than falling into intolerant, anthropomorphic rhetoric. The account is also believable because it is different from others: it was not part of an "official" report and it does not draw on previous representations. Indeed, the painting from Surat does show a slimmer bird than the previous description suggests. Fuller and Hume consider this image "accurate," as the other birds in the painting—and some still survive— are easily recognisable. Taken together with the previous image by Laerle, the painting begins to give us an altered impression of the species.

Examining the Bones

But does physical evidence and recent technology for testing and analysing physical remains support these few isolated early records? This largely depends on the methods used to arrange or produce scientific artefacts and data. In 1865 a Mauritian schoolteacher with an interest in natural history found a great many dodo bones in the Mare aux Songes, a marsh close to the present International Airport in the south of Mauritius. Anatomist and palaeontologist Richard Owen used this fossil material to construct two skeletons. In his first attempt in 1866 he based his reconstruction on Savery's classic image, producing a stooping, bulky structure. This is an indication of how popular illustrations derived from brief or inadequate reports by sailors sometimes influenced the production of nineteenth-century "science."

Owen's second attempt in 1872 was a more upright skeleton that is similar to the few reports of a dodo with long legs and a smaller frame (Figure 11.4).[36] Then,

---

[33]   Hume, "History of the Dodo," 70; Fuller, *Dodo*, 82–7.

[34]   Fuller, *Dodo*, 66.

[35]   Ibid., 65.

[36]   Hume, "History of the Dodo," 85–6.

Fig. 11.3   Drawings of a dodo by Joris Joostensz Laerle, 1601.

Fig. 11.4   Richard Owen's second reconstruction of the dodo (1872) from
              skeletal material found in the Mare Aux Songes, Mauritius. In
              Owen, *Transactions of the Zoological Society of London*, 1872.

in the late twentieth century, zoologist Andrew Kitchener measured dodo bones
held at Cambridge University and the Natural History Museum in London and,
using the mean measurements, reconstructed a long-legged and thin-necked dodo.
He also tested eggshells and sliced bones for cantilever strength and, using four
different methodological processes, he estimated that the bird weighed between
10 and 17 kilograms—much less than the 50 pounds (23 kilograms) minimum
stated by a traveller, Tomas Herbert, in 1634. With these and other factors in mind,
Kitchener eventually came up with a size and shape that largely corresponds to the
dodo image made at Surat and the anonymous sailor's report. He then compared
cantilever strengths of other animals and correlated this with Evertszen's statement
that the dodo "could run very fast." In an article published in *New Scientist* in

1993 Kitchener writes: "for more that 350 years the dodo has been thoroughly misrepresented as plump and immobile. The reality is, however, that in the forests of Mauritius it [sic] was lithe and active." He adds:

> like other Mauritian birds [the dodo] would have undergone a seasonal fat cycle to overcome shortages of food, but never to the extent that those wonderful oil paintings suggest. Sadly, it is from these portraits of the last captive individuals that most people have gained their impressions ... [37]

DNA from dodo bones now supports classification of the dodo as a close relative of the Mascarene Solitaire—an upright, graceful bird quite able to outrun a human in rocky terrain.

The Question of Agency

If dodos could run fast, then why did they become extinct? Consistent with other unique fauna of small islands, birds like the dodo had no natural predators and the major reason for extinction is now believed to be the introduction of exotic animals to Mauritius by Portuguese, Dutch, and British mariners even before 1600—pigs had reached high densities by 1620 and plague proportions by 1700—and extensive habitat destruction and deforestation, such as the harvesting of ebony forests. Direct human hunting—the activity that gave rise to the statement that dodos "allow[ed] themselves to be beaten to death"—was probably limited, especially as the human population during the species' existence averaged fewer than 50 at any time.[38] Turvey and Cheke describe the idea that the dodo's "gross obesity and lack of intelligence" contributed to the species' extinction as a "deeply entrenched misperception" that was widely held in the nineteenth century.[39]

So where does agency lie? The story of the dodo and many other extinct species is complex and the historical evidence remains inconclusive, but it is clear that dodos were not necessarily passive. If the animal had been the centre of Gooding's research, he would have looked at the evidence much more carefully. He may have found scientific and historical support for a different story: one where direct human confrontation with dodos was rare and, when it did occur, dodos sometimes resisted attempts at capture. As Jonathan Balcombe has pointed out earlier in this volume, there is now ample evidence that animals have consciousness and intentionality. In the introduction to their edited collection *Animals and Agency*, human-animal studies scholars Sarah McFarland and Ryan Hediger comment on novel forms of agency expressed in animal sounds and reasoning, and the limitations human language places on any objective judgements in relation to other species. They also note that "when it comes to the mental processes of another being, human or

---

[37]   Kitchener, "Justice at Last," 27.

[38]   Hume, "History of the Dodo," 82.

[39]   Turvey and Cheke, "Dead as a Dodo," 158.

otherwise, there is always a point beyond which each of us as individuals cannot comprehend," and point out that numerous, surprising behaviours of nonhuman animals warrant the title of "agency."[40]

In his essay "Anti-This—Against-That" geographer Chris Wilbert takes a more cautious view, tracing the development of the "modern constitution" that separated the social world from the natural world and the recent rise of debate around the question of animal agency. Wilbert discusses whether nonhuman animals are capable of resistance "in the sense that is implied by some eco-anarchist groups," especially in relation to wild animals. He comments on the high degree of domination and exploitation involved in all kinds of human-animal relationships but notes that, paradoxically, wild animals have been associated with "freedom"— that is, with resistance to capture. These narratives, he feels, "play on, and play up, the disturbing, destructive, qualities of 'wildness.'" Although he seems sceptical, Wilbert admits that judging from some press reports "animals do indeed appear to act purposively." He quotes Tim Cresswell who argues that resistance is "a purposeful action directed against something disliked."[41] Wilbert's conclusion is that it is possible to argue that nonhumans are not so different to humans in regard to actions and that agency is a relational effect: something which emerges out of interactions. Ironically, to support his view Wilbert calls on Gooding's (1992) comment, made in relation to scientific studies of animals, "recalcitrances shape and constrain the development of experimentation: they enable empirical constraints." Wilbert finishes with the idea that nonhuman resistances occur "at the micro-levels of everyday life" and as a result are often missed, ignored, or are unattributed in social and cultural accounts that tend to "attribute agency to *strategic* speakers" (emphasis added).[42]

While these ideas about agency have relevance to stories about the dodo, in many situations where humans and animals interact humans have a greater capacity to effect change, often through indirect agency. For instance, through the mass introduction of new species, the consistent upsetting of biodiversity that favours one or another species, and the widespread alteration or clearing of vegetation. These more subtle agencies can have a snowballing effect and are much harder to pinpoint as causal. Another way humans have exercised profound agency is through representing species like the dodo and writing their histories. How animals are represented clearly matters for, as Steve Baker points out, these representations "have a bearing on shaping 'reality,' and ... the reality can be addressed only through the representations."[43] It is obvious from the history of European experience that dodos and many other animals are "known" much more in representation that in reality. This is true not only in regard to endangered species that are rarely seen today except in the highly selective, stylised, and commodified world of new media, but also in the nineteenth and early twentieth century when colonial animals were represented in engravings and early photographs.

---

[40]   McFarland and Hediger, *Animals and Agency*, 8–9.
[41]   Wilbert, "Anti-This—Against-That," 243–9.
[42]   Ibid., 250–52.
[43]   Baker, *Picturing the Beast* (2001), xvii.

One such animal was the thylacine, an unusual marsupial-carnivore only living on the island of Tasmania, Australia, who was represented as a vicious, wolf-like predator. Images and texts built up a network of inaccurate and misleading "facts" about this shy, pouched animal that was rarely sighted in the wild. Stories of sheep killing were wildly exaggerated and bounties, based on this and other erroneous information, soon wiped out the small, remnant population of thylacines on the island.[44] Other wolf-like animals, such as the Antarctic Wolf or Warrah and the Japanese Wolf, were similarly represented and became extinct.[45] On the other hand, Australian koalas, who are pictured as cuddly bears and celebrated in children's literature, have survived despite small numbers, limited food sources, and threats to their habitat.

While there are many examples of how humans, as the makers of representations, have exercised the most potent agency—the ability to destroy a species—they also have the capacity to write histories "in which animals and humans no longer exist in separate realms; in which nature and culture coincide; and in which animals, not just humans, have shaped the past."[46] Or as Wilbert expresses it, "where the world is viewed as an interrelational field of agency."[47] Only humans have the ability to place animals at the centre of historical narratives and to disassemble anthropocentric stories. While there exists an assumption that humans can only be self-centred, history has shown that brief moments of insight, as well as radical changes in attitude, are possible. The new field of human-animal studies can contribute to a transformation in human relations with animals just as profound as recent events that have brought formerly oppressed human groups and individuals into mainstream society, and even to the pinnacle of political prominence. Perhaps some time in the future a made-over dodo will act as a mascot for the Mauritian Olympic team. In similar vein, Kitchener asks in his article "Justice for the Dodo"—could a slim, active dodo have won the Caucus race in *Alice in Wonderland*?[48] But Carroll's textual representation of this much-maligned species should have the last word. Despite Tenniel's famous illustration, Carroll's dodo is a wise and considerate bird: when asked who has won the Caucus race the dodo replies after a great deal of thought, "*everybody* has won, and *all* must have prizes."[49] As Stephen J. Gould has suggested, if humans embraced this conceptual metaphor, they may find themselves in a better position to face the moral consequences of their actions.[50]

---

[44] For details of how representations contributed to the extinction of the thylacine see Freeman, *Paper Tiger*.

[45] See Flannery, *A Gap in Nature*; Day, *Doomsday Book of Animals*; Knight, "On the Extinction of the Japanese Wolf."

[46] Fudge, "History of Animals."

[47] Wilbert, "Anti-This—Against-That," 251.

[48] Kitchener, "Justice at Last," 27.

[49] Carroll, *Annotated Alice*, 49.

[50] Gould, "Dodo in the Caucus Race," *Natural History*, 33.

Chapter 12
Cetaceans and Sentiment

Philip Armstrong

I

History is moved by emotion, no less than by ideology or economics.

We are accustomed to thinking that social and cultural change occurs because of the impact of ideas whose time has come: this is the humanist view of history. And we are familiar with the alternative theory, that the beliefs and actions of people are largely determined by their material conditions: this is Marxist historiography. We tend to be far less adept, however, at accounting for the role of emotional states in shaping historical phenomena.

One reason is that emotional realities are evanescent and hard to document empirically. Even the most powerful of feelings might leave little record of their passing. In fact, paradoxically, the difficulty of tracking emotional moods might increase with their dimensions. It is easy enough to accept autobiographical or biographical evidence of an individual's emotional life, but it is nearly impossible to verify positively the existence and impact of shared states of feeling dispersed throughout a culture, social group, or population.

There is a second reason for the tendency to underestimate the historical role of states of emotion. Over the last few hundred years, in accordance with the emphasis on reason and empirical observation that is the legacy of the European Enlightenment, attentiveness to emotional realities has come to be associated with subjectivism, lack of scholarly rigour, intellectual weakness—all the epistemological vices that we tend to associate with the term "sentimentalism."

Raymond Williams confronted this historiographical blind spot when he suggested the term "structure of feeling" to refer to those shared states of emotion that operate in specific historical moments and places, as they are influenced by, and as they influence, the surrounding socio-cultural forces and systems. Williams defined the "structure of feeling" as a "lived" or "practical consciousness" of meanings and values, prior to their explicit articulation, definition, classification, or rationalisation in fixed or official ideologies: "it is a kind of feeling and thinking which is indeed social and material but each in an embryonic phase before it can become fully articulate and defined exchange."[1]

Because societies so often express their collective emotional dispositions in relation to other species, interactions between humans and other animals provide

---

[1]   Williams, *Marxism and Literature*, 130–31.

one crucial source of data on the structures of feeling that move history along. Hence, for example, the provenance of taken-for-granted structures of feeling such as compassion, humaneness, sympathy, and sentiment cannot be understood properly without reference to their conceptualisation vis-à-vis nonhuman animals during the modern period.[2]

II

For Raymond Williams, literature was the best source of data for studying the way structures of feeling reflect and shape social histories. But of course any cultural practice, text, product, site, or event might be just as fruitful. For instance, "human interest" stories on TV news, because they are specifically designed to evoke or appeal to the feelings of a wide audience, offer a rich focus for this kind of inquiry—and not least because, very often, they are actually "animal interest" stories.

In 2006 New Zealand's TV3 News featured a story of this kind. The piece starts with footage of children jumping off a jetty into a blue-green sea at Opononi, a small settlement on New Zealand's Hokianga Harbour. The voiceover begins, "Another Opononi summer, where the kids play in the water. But there was a time, a magical time, when someone else played with them, and made their town famous. It was the summer of Opo, in the mid-fifties." At this point colour fades to black and white as the contemporary scene is replaced by film footage from the 1950s. A dolphin is chasing a beach ball around amongst a crowd of adults and children standing knee-deep in the shallows. The reporter describes how, over 11 weeks in the summer of 1955–56, a wild dolphin visited Opononi Beach almost every day, swimming and playing with the increasingly large numbers of people who came to see her. "Opo was a two-year-old bottlenosed dolphin," says the voiceover, "who loved and trusted human company." The story is intercut with short interviews with locals who were children at the time, and accompanied by archival film footage and photos. Included are extraordinary images, moving and still, of the dolphin ducking between people's legs, bouncing balls and beer-bottles off her nose, and allowing small children to be placed on her back.[3]

Opo's visit became embedded in New Zealand popular culture with extraordinary speed, potency, and durability. For half a century, in many different genres, it has been told and retold. The events were relayed to the New Zealand public via extensive newspaper and radio reporting at the time, and the story was picked up by many overseas papers as well. The first extensive eyewitness account was written by Piwai Toi, a local farmer, and published in *Te Ao Hou*, the Māori Affairs Department journal, in 1958.[4] But perhaps the longevity of the

---

[2]   For a fuller discussion of this claim see Armstrong, *What Animals Mean*.

[3]   McNeill, "Opo Fifty Years On."

[4]   Toi, "Opo the Gay Dolphin."

story has more to do with the way it entered so quickly into various genres of popular culture. Almost immediately the events in Opononi generated a hit song, "Opo the Friendly Dolphin," which can be heard in the background of the TV3 story. Around the same time there was a children's book, *Opo the Gay Dolphin* by Avis Acres, which is still in print today. Prominent film-maker Rudall Hayward was sent to Opononi;[5] his remarkable footage is used in the TV3 piece along with some of the stills taken by Eric Lee-Johnson, part of a photographic sequence that became justly famous.[6]

Extensive retellings of the Opo story were forthcoming throughout the ensuing decades. In 1960, cultural historian Antony Alpers produced a book-length treatment of the story which placed Opo in the context of dolphin-human friendships in ancient times.[7] A novel based on the events was published by well-known New Zealand writer Maurice Shadbolt in 1969, featuring a striking cover painting by artist Michael Smither, who was so struck by the theme he continued to paint Opo-inspired images for the next decade.[8] Another documentary, *Opononi: The Town that Lost a Miracle,* was made in 1972 by firebrand Māori film-maker Barry Barclay. Meanwhile Opo's place in New Zealand children's culture was confirmed by a Little Golden Book version of the tale, *Opo the Happy Dolphin*, written by Julia Graham in 1979. The story has continued to circulate in the new millennium. *The People-Faces*, a novel by Lisa Cherrington published in 2002, uses the memory of the dolphin's close relationship with one particular child as the emotional fulcrum for a novel about mental illness and Māori-Pākehā relations in contemporary Northland. And in 2006, 50 years after the original events, popular Kiwi singer-songwriter Don McGlashan had a hit with a song about Opo entitled "Miracle Sun."

Most strikingly, in all these different versions (with only one significant exception, which I'll come back to later) the Opo story has retained certain key elements unchanged. The first is the animal's particular connection with children, and with one child in particular, a 13-year-old girl called Jill Baker. Piwai Toi wrote that Opo

> was really and truly a children's playmate. Although she played with grownups she was really at her charming best with a crowd of children swimming and wading. I have seen her swimming amongst children almost begging to be petted. She had an uncanny knack of finding out those who were gentle among the young admirers, and keeping away from the rougher elements.[9]

---

5 Movietone News.
6 Lee-Johnson and Lee-Johnson, *Opo the Hokianga Dolphin*.
7 Alpers, *A Book of Dolphins*.
8 Shadbolt, *This Summer's Dolphin*.
9 Toi, "Opo the Gay Dolphin," 22.

Lonely animal and lonely child make friends: a classic sentimental trope. Literary historian Philip Fisher suggests that "the typical objects of sentimental compassion are the prisoner, the madman, the child, the very old, the animal, and the slave."[10]

Another trait of sentimental narrative, according to Fisher, is the experimental extension of selfhood, subjectivity, and of "normal states of primary feeling" to those "from whom they have been previously withheld."[11] This is also the second feature of the Opo story that remains consistent throughout its 50-year history. All the retellings accord the dolphin two attributes of subjectivity: agency and personality. A conventional prejudice about wild animals, that they only operate on instinct, is suspended and replaced with the repeated assertion that this animal made an intentional choice to consort with humans over an extended period of time. Opononi publican Ian McKenzie, who played with Opo as a child, sums this up in the TV3 news story: "it was her choice to make friends with human beings and she didn't hold back."[12] Moreover, as shown by the remarks of Piwai Toi cited above, Opo was also ascribed a charismatic personality typified by gentleness combined with exuberance, a benign sense of fun embodied in the dolphin's apparently smiling face.

A third important element common to the many Opo narratives involves the operation of a special period of grace, attributable to the presence of the animal, which suspends the competitiveness, aggression, violence, and grief of everyday human life. In a detail that became inextricably associated with the Opo legend, Piwai Toi describes how "[s]ome people got so excited when they saw Opo that they went into the water fully clothed just to touch her."[13] This moment is wonderfully captured in one of Lee-Johnson's photos, which shows a crowd of men, women and children standing thigh-deep in luminous water, gazing seaward in apparent rapture (Figure 12.1). As Alpers put it a few years later, "[o]ften it must have suggested some Biblical scene, with simple believers crowding to touch the garment of a Holy prophet and gain redemption."[14]

Furthermore, in a statement confirmed by every other source I have seen, Toi asserted the following:

> With the record traffic on the roads I never heard of a single motoring accident in coming to or returning to Opononi. As for swimmers there were easily over a hundred young and old in the water but there was not a single drowning fatality. With all these people coming during the weekends, Saturdays in particular when up to 1500 people were jammed on the beach, there was no case of drunkenness, fights or arguments.[15]

---

[10]   Fisher, *Hard Facts*, 99.

[11]   Ibid., 98.

[12]   Quoted in McNeill, "Opo Fifty Years On."

[13]   Toi, "Opo the Gay Dolphin," 22.

[14]   Alpers, *A Book of Dolphins*, 136.

[15]   Toi, "Opo the Gay Dolphin," 22.

Fig. 12.1   *Dolphin Daze*. Eric Lee-Johnson, Opononi, 1956. Te Papa Tongarewa
Museum of New Zealand. Used with permission.

For the duration of the dolphins' visit, as Alpers comments wryly, "Saturdays, for
all the crowds and all the beer, were not as New Zealand Saturdays so often are."
Instead it was as if "[o]n this mass of sun-burned, jostling humanity, the gentle
dolphin had the effect of a benediction ..."[16]

The reconciliatory atmosphere also extended to Māori and Pākehā relations.
The name of the harbour the dolphin visited, "Hokianga," means "returning-place,"
so called because it is believed to be the landing-site of Kupe, the Polynesian
discoverer of Aotearoa, on his return voyage to the north. Toi cites kaumātua
(Māori elder) Hohepa Heperi's assertion that "Opo is the fish of peace, a legacy
from Kupe."[17] In other accounts it is said that Opononi Māori believed the dolphin

---

[16]   Alpers, *A Book of Dolphins*, 132.
[17]   Toi, "Opo the Gay Dolphin," 22.

to be Kupe's incarnation or at least his messenger. One kaumātua announced that "Opo's mission in life is to bring together all people, of all races, in peace and friendship."[18] When Opo died, her funeral included extensive Māori ceremonials as well a message of condolence from the Governor-General.

The final characteristic of the Opo story which I want to highlight concerns the time-scheme that typifies sentimental tragedy. According to Fisher, sentimentalism tends to evoke two kinds of temporality.[19] The first is the moment at which action comes just too late to prevent a tragedy. This how McNeill portrays Opo's death in his commentary for the TV3 item:

> But the sunny days of Opo's summer were short-lived. There were already fears for her safety, and on March the eighth 1956 dolphin protection legislation became law. But it was all too late. A day later she was found dead, jammed between rocks at a point across the harbour. The cause of her death has never been determined, but many believe Opo was accidentally killed by someone using gelignite to catch fish.[20]

The second time-scheme Fisher associates with sentimental narrative is the event "in the deep past" which has "left irreversible damage."[21] This is how the Opo story is recalled now; hence, McNeill finishes his news item as follows:

> The summer of 1955 and 56 was a time of innocence, a time when the days seemed warmer, without the UV rays, a time when people seemed friendlier; but when Opo was found dead in the rocks, it was over. The summer had ended, and there's never been another one like it since.[22]

The TV3 item thus concludes by inviting a classic sentimental nostalgia for the golden days of childhood—so often associated in New Zealand with the 1950s. Like most forms of shared nostalgia, however, this structure of feeling relates closely to contemporary anxieties. When the 1950s sun is said (surely inaccurately) to be "without the UV rays," we are reminded of current health and environmental concerns (melanoma, the depletion of the ozone layer). Even the friendliness of the people who shared Opo's beach recalls various present-day problems: their lack of drunkenness highlights our struggle with substance abuse; their lack of aggression contrasts with our rates of violent crime; their safe driving reminds us of our road toll. Of course it is not suggested that Opo's death is the cause of these contemporary ills, but the implication (which operates according to the associative logic typical of sentimentalism) is that the tragic termination of Opo's visits symbolised our loss of innocence as a society.

18　Quoted in Shadbolt, "Opo," 185–6.
19　Fisher, *Hard Facts*, 108.
20　McNeill, "Opo Fifty Years On." Gelignite is a type of dynamite.
21　Fisher, *Hard Facts*, 108.
22　McNeill, "Opo Fifty Years On."

The same mood—that of nostalgia for an irrecoverable innocence, and of sorrow for an irremediable tragedy—is evoked by McGlashan's "Miracle Sun." The song's melody moves from repetitive, meditative verses into a soaring and plaintive chorus, and its lyrics reflect the same tropes described above:

> They say when people get to the water
> they don't undress, they just keep on walking out from the beach.
> They say men weep when they see the dolphin.
> He[23] comes in close and does tricks for the children just out of reach.
> They say he lets you go right up and touch him.
> They say he lets you ride on his back, all slippery and gold.
> Miracle Sun, shine on the water, shine in my heart.[24]

III

The characteristics of sentimentalism I have borrowed from Philip Fisher derive from his discussion of Charles Dickens's *Hard Times* and Harriet Beecher Stowe's *Uncle Tom's Cabin*. It might be asked why, in considering mid-twentieth century events and their subsequent reworkings, I have chosen to draw on ideas formulated in response to nineteenth-century sentimental narratives. The answer is that throughout the twentieth century sentimentalism was treated by scholars and researchers of all sorts, from scientists to literary critics, with almost universal contempt. The overwhelming tendency in critical or theoretical discourse written over the last century is to dismiss sentimentalism, or to refer to it in an unthinkingly pejorative manner. As Fisher comments, today the term "sentimental" is widely used to denote "false consciousness of a particularly contemptuous kind," or to describe cultural effects that are "cheap, self-flattering, idealizing, and deliberately dishonest."[25]

Yet this disdain for sentiment is a "peculiarly twentieth-century mistake."[26] During the eighteenth and much of the nineteenth centuries, the sentimental structure of feeling was considered vital to Western aesthetic and philosophical discourse. The development of one's capacity for sentiment, characterised as a heightened sensitivity to the experience and especially the suffering of others, was thought a most appropriate means to achieve both private and public virtue, and the main function of the arts was to achieve this disposition. However, the end of the nineteenth century and the start of the twentieth saw the emergence of an

---

[23] While she was alive, Opo was believed to be a male; it was only after her death that her sex was determined. Hence the first popular song about her, "Opo the Friendly Dolphin," which was recorded before her death, referred to the dolphin as "he"; McGlashan does the same, perhaps in reference to the earlier song.

[24] McGlashan, "Miracle Sun."

[25] Fisher, *Hard Facts*, 92.

[26] Ibid.

aesthetic and critical "campaign to purge the [arts], generally by means of irony, of [their] sentimental texture."[27] For the modernists of the first half of the twentieth century, Victorian sentimentalism represented everything they wished to leave behind: the consolation of miraculous escapes, the mordant sorrow of drawn-out tragedies, clichéd forms of affect and aesthetics, and most of all the consolations and complacencies of the bourgeois family.

The emergence of a cultural and intellectual contempt for sentimentalism—especially in relation to nonhuman animals—had everything to do with the simultaneous rise of positivist science and industrial capitalism. In both of these domains, compassionate feelings for nonhuman animals in particular were perceived as obstacles, either to progress or profit. The industrialisation of animal farming and slaughter, proceeding apace in locations such as Chicago, required a radically dispassionate view of nonhuman animals.[28] At the same time, the increasingly strict intellectual separation between the humanities and the sciences was working hard to arrogate to the latter domain sole expertise in regard to animals, and indeed all of nonhuman nature. The official triumph of the culture of science over the culture of sentiment can be seen in turn-of-the-century conflicts such as the "Old Brown Dog" riots in Battersea, where a militant and powerful antivivisection movement—largely female and working class in membership—was finally defeated by the middle-class and overwhelmingly male-dominated London Medical School.[29] For the next several decades—at least until the 1970s—the operations of commerce, industry, and science would be more or less invincible to challenges based on human sympathy for the suffering of animals—precisely because such challenges could be assigned to the increasingly devalued category of the merely sentimental. During much of this period, for those charged with the task of considering animals at a professional level—biologists, or those in agribusiness, or literary and cultural scholars—the dominion of science, industrial capitalism, and modernism made sentimentalism something between a crime and a sickness.

The manifestation of this anti-sentimentalist hegemony in the arts is well illustrated by what happened when prominent New Zealand novelist Maurice Shadbolt undertook to write a literary version of the Opo event. His novel *This Summer's Dolphin* relocates the story to a fictional offshore island in the late sixties, and although the general contours of the narrative are the same, the novel is shaped by Shadbolt's determination to divest the event of its sentimental pathos and give it instead a bold, savage, modernist authenticity.[30] Accordingly, in his novel the dolphin is not gentle at all, but rough and dangerous in play; a sublime rather than sentimental presence that provides a kind of exhilaration bordering on

---

27   Ibid., 92–3.

28   As notoriously portrayed in Upton Sinclair's *The Jungle*. For further discussion of the impact of science and industrial capitalism on human-animal relations see Armstrong, *What Animals Mean*.

29   Lansbury, *The Old Brown Dog*.

30   Shadbolt, *This Summer's Dolphin*.

terror and annihilation. Indeed one of the novel's characters, a university professor who sees the dolphin as an invitation for re-union with the spirit of nature, walks fully clothed into the ocean towards the circling fin, only to find that his inadequate training in marine zoology has betrayed him, since the fin turns out to belong to a large shark. His remains are washed up on the beach the next day.

Similarly, instead of choosing for special companionship a gentle, lonely 13-year-old girl, Motu (the dolphin in Shadbolt's novel) bonds with David, a spectacularly troubled 18-year-old boy, traumatised by having shot his twin sister dead the year before in a game of dare that went too far. The townsfolk are also far less benevolent in the novel than they are in other versions of the Opo legend. Shadbolt's imagined community resolves to confine Motu in a concrete pool in order to retain tourist interest in the town. Consequently (since the novel was written in 1968) a contingent of hippies arrives from the mainland to defend the dolphin's freedom and a violent confrontation with police ensues. Finally one of the townsfolk, a returned serviceman disgusted by the disorderly state of contemporary society, shoots the dolphin from the beach. In keeping with the modernist aesthetics of alienation, *This Summer's Dolphin* includes a graphic, drawn-out, and cruel depiction of the animal's death. Finding Motu washed up on the beach, dying from the gunshot wound, the boy attempts to return the dolphin to the water and becomes so desperate that a woman who has befriended him takes a rock and smashes the animal's skull to pieces; the boy then attacks her viciously in turn.

Although *This Summer's Dolphin* is a powerful novel, not even Shadbolt claimed that it represented a more accurate version of the story. Shadbolt was himself resident in Opononi during that summer of 1955–56, and witnessed Opo's visits first-hand. Some years after publishing his novel, he wrote a non-fiction account of the actual events, which in every respect agrees with the sentimental content and tone of all other manifestations of the Opo story—and therefore contrasts markedly with his modernist version in *This Summer's Dolphin*. Thus, in his eye-witness account, Shadbolt agrees with other documentarians that "no one was ever hurt by Opo."[31] Describing Opo's real-life friendship with Jill Baker, he writes: "A lonely girl, an only child, she thought the dolphin lonely too. Their friendship, in all that human turbulence became the more moving."[32] And reflecting on the reaction of the Opononi townsfolk and the visiting tourists to their cetacean companion, he insists that, far from causing the kind of self-interest and conflict he describes in his novel, "... Opo had the effect of a benediction on the growing throng of humanity ... Opo offered more than enough to the mystical; the event had all the atmosphere of a miracle."[33] In the final words of this account he even describes "that half-lost summer" as "the only fairy-tale I am ever likely

---

[31] Shadbolt, "Opo," 184.

[32] Ibid., 184–5.

[33] Ibid., 185–6.

to live in my life."[34] The contrast between Shadbolt's two accounts of Opo—the sentimental but true-to-life recollection, and the artificial exercise in modernist realism—illustrates perfectly the conflict between sentimentality and modernism as described by Fisher.

IV

Perhaps the reason Shadbolt felt able to reconfirm the popular, sentimental version of the Opo story a decade after his novel was published was that during the seventies such narratives were achieving an increasingly widespread and significant place in public discourse and action. In this respect, stories like that of Opo demonstrate another aspect of the mistake twentieth-century thought makes in relation to sentimentalism, in considering it not only a failure of taste, but a form of failed politics: at best ineffectual, and at worst conservative or reactionary. In spite of these prejudices against it, sentimentalism has often had a very strong relationship with progressive politics. Fisher describes the function of nineteenth-century sentimental narrative as "a radical form of popular transformation" which proved capable of "massing small patterns of feeling in entirely new directions." Sentimentalism, deployed in this way, "trains and explicates new forms of feeling," offering "a necessary practice for a transformation of moral life that is approaching."[35]

It is my contention that, in the second half of the twentieth century, experiences like the Opo event acted in precisely this way. Such stories were circulated over and over again, I suggest, because they provided concentrated, vivid, highly charged emblems for radical transformations of popular cultural affect that were occurring at the time. This point can be illustrated by considering the most famous and clichéd of Lee-Johnson's images: the one of a smiling Opo embraced by Mrs Goodson, a local schoolteacher (Figure 12.2). Such is our familiarity with this kind of image nowadays that it is very hard to imagine how remarkable it must have seemed at the time. But in the mid-fifties only mariners would have had any familiarity with dolphins. The aqualung was not yet widely available; television did not reach the country until 1960; the first popular imagery of dolphins—the series *Flipper*, for example—did not air until the mid-sixties. Nor was there any significant public structure of feeling in favour of cetaceans at work during the mid-fifties; New Zealand was still a decade away (and Australia more than two decades away) from cessation of commercial whaling.

Sometimes the clichéd appearance of sentimental images or narratives indicates the thoroughness with which they have permeated the public imaginary. The staleness we have come to associate with political conservatism might actually be the mark of an enduring achievement. Such is the case with the dolphin smile. For it was during the sixties and seventies that dolphins became visible—indeed

---

34   Ibid., 189.
35   Fisher, *Hard Facts*, 18–20.

Fig. 12.2   *Mrs Goodson and Opo.* Eric Lee-Johnson, Opononi, 1956. Te Papa
Tongarewa Museum of New Zealand. Used with permission.

ubiquitous—as icons of humans' lost relation to nature and as ambassadors of a
planet in crisis. In their book *Cosmodolphins*, Mette Bryld and Nina Lykke assert
that 1960 was the year in which the dolphin, "[m]ysteriously smiling ... entered
the American imagination ...."[36] Up until then, they assert,

> very little research had been done on cetaceans. Not until that year did the
> echo-locating or sonar abilities of dolphins become an established scientific
> fact ... Because of their extra-terrestrial environment, these sea mammals were
> extremely difficult to study, and of the Marinelands or oceanaria that became so
> widespread later on there were at that time only a couple in the USA.[37]

[36]   Bryld and Lykke, *Cosmodolphins*, 83.
[37]   Ibid., 186. Bryld and Lykke capture the suddenness and intensity of the mid-
twentieth century's new fascination with marine life and dolphins in particular, but the
choice of 1960 as the key date is open to debate: see for example Gregg Mitman's account of
the developing exploitation of dolphins for entertainment, biological research, and military
deployment at William Douglas Burden's Marine Studios in Florida during the 1940s and
50s (Reel Nature, 157–79). In New Zealand, however, there were no captive cetaceans until

Quite suddenly, however, within a few years, dolphins and whales began to gain enormous public popularity as "harbinger[s] of a New Age."[38] More and more sentimental accounts began to emerge of charged encounters between humans and cetaceans. And these stories and images were encapsulations of a major change in popular ideology—that challenge to modernity's mastery over nature which we now associate with the environmental and animal advocacy movements.

Placed in context, then, the Opo event becomes a vivid instance of new and culturally significant kinds of emotional connection that were occurring at the time between humans and animals, and especially between humans and cetaceans. Antony Alpers anticipated this when, in portraying the mood that permeated Opononi during the dolphin's visit, he described an atmosphere of reconciliation extending to interspecies relationships:

> There was such an overflow of ... friendly feelings that it seemed the crowds were composed of people wanting to be forgiven for something—for the unkindness, perhaps, that humans generally do to animals in the wild. The dolphin, who never once snapped at a hand, seemed to offer forgiveness to all.[39]

The Opo narrative, then, exemplifies the kind of cultural event that "arouses and excites action toward that part of the public future that is still open to decision and alternatives ..."[40] It does so by inviting its audience, through sentimental effects, to "enter what are only temporarily exotic configurations of experience as a necessary practice for a transformation of moral life that is approaching."[41] To demonstrate this interpretation more explicitly, I want to end by discussing a second Opo-like event.

During the mid-sixties a New Zealander called Paul Spong—a graduate of Canterbury University, subsequently trained in neuropsychology at UCLA—was hired by the University of British Columbia to undertake behavioural research on Skana, a captive *Orcinus orca* at the Vancouver Public Aquarium. Attempting to test the orca's visual acuity, Spong was nonplussed by a sudden and complete turnaround in his subject's performance. After weeks of scoring almost a 100 per cent, Skana suddenly got "exactly the wrong answer eighty-three times in a row." Recognising that these results could not be explained as mere mistakes, Spong began to suspect that Skana, having learnt how to pass the test, was giving the

---

1965, when two common dolphins were captured for Napier Marineland by Frank Robson. By then, the popular image of the dolphin as a marine version of "man's best friend" was well established in the United States, not least by films and TV shows like MGM's *Flipper* (1963; see Mitman, Reel Nature, 177).

[38]   Ibid., 183.
[39]   Alpers, *A Book of Dolphins*, 137.
[40]   Fisher, *Hard Facts*, 18.
[41]   Ibid., 19–20.

wrong answers on purpose. As his wife Linda put it, "[i]s there a Latin phrase for animal rebellion against scientists?"[42]

One day, as Spong sat dangling his legs in her pool, Skana swam up and raked his feet with her teeth, causing him—not unnaturally—to jerk them out of the water. She repeated this behaviour several times, until Spong decided to leave his feet where they were. As soon as he stopped reacting in fright, Skana gave up the behaviour. As Spong saw it, this was Skana experimenting with him.

Beginning to recognise not only that Skana was more intelligent than his first experiments had supposed, but that she was an active subject rather than a passive object in what he was doing, Spong began a series of more unconventional experiments. As with Opo, the cetacean's agency, rather than that of the human, began to define the encounter. At the same time Spong, who had no prior interest in cetaceans and only took the UBC job as a career move, began to develop a powerful emotional attachment to the isolated whale, who had been caught from the wild a few years before. This led to a famous moment of sentimental connection. He would stand at the side of the pool playing his flute to her, and she would listen and at times seem to reply.

Here again, then, the key elements of the sentimental narrative are at work. Skana is both an animal and a prisoner, two of Fisher's categories of likely subjects for sentimentalism. The story represents the experimental—in every sense of the word—extension of subjectivity, of affect and intelligence to this being who was formerly denied it. There is a miraculous but poignantly brief period of grace during which Spong plays his flute and Skana responds, and they learn to trust each other. Then Spong is sacked from his job for publicly calling for the whale to be freed back into the wild. There is a tragic, just-too-late time-scheme: in 1980, just as Spong's long campaign to free her comes close to realisation, Skana dies in captivity and her body is sold for dog food by the Vancouver Aquarium.

The broader political outcome of this sentimental narrative is one of the most familiar historical phenomena of our time. Spong went on to become a researcher of wild orca, and a campaigner against keeping cetaceans in captivity. Then, in 1973, he persuaded a group of Canadian activists, who until that time had been concentrating on protesting against nuclear arms, to begin a campaign to "Save the Whales." And it was this campaign, and especially the famous confrontation on the high seas with Russian whaling ships during the IWC meeting in 1975—a moment Spong helped orchestrate—that resulted in the emergence of this organisation at the forefront of the environmental movement. The organisation was of course Greenpeace.

Paul Spong's sentimental engagement with Skana, then, was a formative inspiration for the course taken by the contemporary environmental movement. While trying to persuade the Greenpeace founders to get involved in the anti-whaling campaign, Spong took Bob Hunter, one of leaders of the group, to meet Skana in her pool. When Hunter leant down to meet the whale, Skana opened

---

[42]   Weyler, *Song of the Whale*, 6–7, 19, 22.

her jaws and gently enclosed Hunter's head between her teeth. Even Spong was alarmed. "One moment I felt more fear than I've ever felt in my life," Hunter told Spong after this experience, "then the next moment I felt a shower of absolute trust."[43] Hunter, too, came away from the encounter committed to the Save the Whales campaign.

Stories like those of Opo and Skana show that, as Steve Baker once put it, "[s]entimentality *matters*."[44] It matters because of its everlasting popularity, which gives it power, but also because sentimental effects may mark the point at which a radical transformation of cultural feeling has taken place—or is about to take place. And so for those who take up the challenge of considering animals, I suspect sentiment might turn out to matter quite a bit.

---

43   Weyler, *Song of the Whale*, 144.
44   Baker, *The Postmodern Animal*, 177.

Chapter 13
Zones of Contagion:
The Singapore Body Politic
and the Body of the Street-Cat[1]

Lucy Davis

The outbreak of SARS (Severe Acute Respiratory Syndrome) in early 2003 provided the Singapore government with an opportunity to renew a historic obsession with hygiene, eugenics, and surveillance. When the virus hit Euro-America, East Asians were racially profiled and designated in the media and by immigration authorities as potential SARS "super-spreaders." In Singapore the fear, economic uncertainty, and shame brought on by SARS, combined with a need to display utmost efficiency, was displaced onto the body of the cat. One emergency action undertaken was a nationwide cull of street-cats, leading to a rare public confrontation between civil society groups and the state.

This essay has two stories rubbing up against each other. The first, "SARS, Cats, and the Singapore Body Politic," recounts how human invocations of the street-cat during and following the "SARS cat-culling incident" (as it has since been called) have inflected a multitude of conflicting gendered, raced, classed, and specied projections. The second, "Night Feeders and the Body of the Cat," winds through this contested territory with a series of human-cat entanglements[2] that took place during one night of street-cat feeding in a public housing estate. The first story is printed in roman type, the second is printed in italics.

SARS, Cats, and the Singapore Body Politic

There are an estimated sixty to eighty thousand street-cats in the city-state of Singapore.[3] One of the most visible icons of urban animal life, cats frequent open-air hawker centres, school canteens, the edges of construction sites, and the

[1] Many thanks to Lee Weng Choy, Sharon Siddique, Ray Langenbach, Adrian Franklin, and Yu-Mei Balasingamchow for input on drafts and edits of this chapter, and to Paul Rae for friendship and provocation.

[2] Much of this essay is informed by Donna Haraway's *When Species Meet*. Haraway considers zones of human-animal relationships as "world-making entanglements [or] contact zones" (4).

[3] This is an unconfirmed government estimate. From mail correspondence with Dawn Kua, Director of Operations, Cat Welfare Society, 2007.

ground-floor areas of government Housing Development Board (HDB) blocks, in which 85 per cent of the human population reside and where they come into contact with highly organised and vocal networks of cat-feeders. One reason why human-cat relations are such a common and contested feature of Singapore public space is because keeping cats in HDB flats is banned. Small dogs are permitted, but cats are deemed too noisy and impossible to maintain inside a flat.

In early 2003, cat feeders informed Singapore's main animal advocacy groups that cats were being captured by pest-control companies. This came as a surprise as the groups had formerly been engaged in joint Trap-Neuter-Release schemes with town councils, whereby cats were permitted to remain in public areas as long as they had been sterilised, with one ear clipped to indicate their neutered status.

One trigger for the cull was a misrecognition of species. The SARS virus was claimed (although this has been challenged)[4] to originate from a palm civet or "civet cat" consumed in an exotic meat restaurant in Guangdong, southern China. Although the civet is not related to the domestic cat according to current taxonomies, rumours proliferated in Singapore that domestic cats were spreading SARS, leading to a spike in acts of cruelty and a government attempt to remove cats from public areas.[5]

After trying unsuccessfully to engage the Ministries of Health and Environment in dialogue, animal advocacy groups convened a press conference in May 2003 where they revealed that a large-scale cull was occurring. This was the first time in recent memory that a press conference was held in response to Singapore government policy without the government being informed beforehand. The next day in *The Straits Times* the government was forced to admit the cull.[6]

Although many human cat advocates are male, media reports during the cull focused on female activists. What is interesting here is that a media profiling of "cat women" corresponded to the same constituency that were so problematised during the infamous "Graduate Mother Scheme" of 1983. That government scheme gave financial incentives for graduate women to produce more children and similar incentives for low-education, low-income families who agreed not to. At stake in the "crisis of patriarchal reproduction"[7] were two groups of deviant women. First, university-educated single women (subtext: Chinese) were pronounced to be neither marrying quickly enough nor dutifully producing enough children. Secondly, uneducated working-class married women (subtext: Malay and Indian) were producing too many purportedly less productive offspring. The scheme was phased out after a rare public outcry from middle-class and feminist groups, and a (relatively) poor showing by the governing People's Action Party in the

---

[4]　Skatssoon, "Civet Cats."

[5]　This was much later rationalised in terms of general hygiene, not directly related to SARS and the unfortunate palm civet.

[6]　"AVA Admits." See also "Nuisance Stray Cats."

[7]　Heng and Devan, "State Fatherhood," 344.

1984 elections.[8] Twenty years on however, the gender, race, and class axes of this episode inflected public discourse on "cat women"—an uneasy conflation of the animal and the woman.

Night Feeders and the Body of the Cat

I am a lapsed cat woman. I moved away in 2004 from the densely populated area in which I used to enjoy nightly communions of varying intensities with 10 to 15 feline individuals.[9] It was therefore with some nostalgia that I arranged to follow Jane, whom I contacted via one of the rash of Singapore feline networks that have emerged in cyberspace, and Daisy, whom Jane identified as a particularly dedicated cat-feeder, on a night feeding tour.

So there were three human females that night, all with different degrees of engagement with cats. I am a 40-year-old single Caucasian Singapore permanent resident and artist/academic with three to five street-cats sharing my home at any given time. Jane is a married Chinese white-collar administrator in her thirties with 10 former street-cats in her flat in the block opposite Daisy's. Jane supports cat-feeders, negotiating with town councils and subsidising food, vet, and sterilisation bills. But she has passed the responsibility of street-cat feeding to Daisy, a single Chinese telemarketer of about 50 who is a prominent feature of the estate. Daisy has nightly contact with around 30 cats, in several different human-cat contact zones which run along public pathways, storm drains, and between housing blocks. She is a highly vocal personality who enthusiastically performed various refractions of the "crazy cat lady"—at times embracing of the moniker in an ironic, camp manner ("Here comes the cat woman!"), then switching and disavowing the label: "Those bastard devils say I am a crazy cat lady but I am just doing my best for the creatures of Lord Jesus!" Jane acted as a translator of Daisy's behaviour and qualifier of Daisy's more outrageous statements, just as Daisy "translated" the behaviour of the cats.

***

The evening commenced in Daisy's two-room corner flat on the tenth floor, where three of her "babies" currently reside. Daisy uses the term "babies" for cats with whom she feels a certain individuated affinity—often those who have spent time in her home after an injury or operation.

---

[8]   The government's share of votes in the 1984 elections fell by more than 10 per cent to below 70 per cent—the biggest fall and the lowest percentage for the party since 1963. Speculation was that the Graduate Mother Scheme was one reason for the fall in support.

[9]   When I moved I arranged for other night feeders to extend their zone of street-cat connections to include mine. They still send SMS updates about how "Panther is evading sterilization" or "Anak still hisses and purrs at the same time."

None of the three cats in Daisy's flat appeared to be comfortable with human strangers. We only met two as the third was hiding. There was a female white kitten with an infected eye who sniffed our feet but who rapidly bat-batted my hand, claws extended, when I bent down to her. An older, large-boned, dark brown male with blue eyes was hidden behind the layers of plush toys, hanging plants, garlands, and paraphernalia that line Daisy's living spaces. "Sorry, my friends say I live in a jungle!" she apologised.

Daisy assured us that when they were alone a different culture prevailed, which is typical for semi-feral cats, who develop intimate connections with select people but still fear humans as a whole. Daisy told us how her everyday routines were punctuated by feline interventions: how they would all sleep on her bed, waking her by jumping on her head, batting her nose, licking her; how human and felines together would play with her other pet—a large white rabbit—crouching, ambushing, and chasing each other round the flat.

***

Gilles Deleuze and Felix Guattari's much-cited chapter "Becoming-Intense, Becoming-Animal" has little time for everyday, emotional relations between real life pet owners and their "little cats or dogs," regarding such "sentimental" relationships as "invitations to regress into narcissistic contemplation" and Oedipal fantasies.[10] However, while Daisy may well be projecting Oedipal fantasies onto her "babies" and at times refers to street-cats as cutesy incarnations of the soft toys that line her flat, it is doubtful that this articulation has much interpellative effect upon the cats in question.

Moreover, as I have argued elsewhere,[11] when I exclaim to my black street-cat Toby, "You are such a beautiful boy!" I may be calling out to him through the ideological frameworks of human familiarity, but I also understand that he is not "my boy," I do know that he is a cat. And our physical communion at such moments, when Toby very slowly blinks and kneads his claws, or stretches his

---

[10]   Deleuze and Guattari distinguish between three kinds of relating to animals, which they also suggest might conceivably occur with the same creature. Lowest in their hierarchy is a relating to animals as family pets "… sentimental, Oedipal animals each with its petty history, 'my' cat, 'my' dog. These animals invite us to regress, draw us into narcissistic contemplation, and they are the only kind of animal psychoanalysis understands, the better to discover a daddy, a mommy, a little brother behind them (when psychoanalysis talks about animals, animals learn to laugh): *anyone who likes cats or dogs is a fool*." They list secondly a relating to animals as state symbols, mythological characteristics, and archetypes. They see greatest progressive potential in the third form of relating—to the pack or swarm: "Becoming-animal" is for Deleuze and Guattari a demonic, transcendent process whereby in certain moments humans get outside themselves via a channelling of the necessarily unformed (and thereby progressive) affective energy of the "pack." "Becoming Intense," 240–42.

[11]   Davis, "About FOCAS 6."

forepaws and arches his rump, or emits a low, lilting prrrowl and takes off to furiously scratch-scent a tree stump or sofa, has of course little to do with the words I am saying. It has very probably something to do with the tenor or magnitude of my intonation,[12] possibly with my body as I clumsily bend to approximate a world where four paws connect with warm concrete or cool tile, possibly also with an unformed intensity behind my utterance that cannot be reduced to a definitive register of human emotion or articulation—something that Deleuze and Guattari might call "affect."[13]

Donna Haraway in *When Species Meet* and Vicki Hearne in *Adam's Task* have embraced individuated relations between humans and "companion species" and explored how, for example, training and agility collaborations with dogs and horses co-choreograph everyday worlds of interdependent possibility—or what Haraway calls "becomings with." Haraway distinguishes an always-in-process, co-producing "companion species" from the historically situated "companion animal." For companion species, "the partners do not precede their relating; all that is, is the fruit of becoming with." Haraway's "becoming with" is to be distinguished from Deleuze and Guattari's "becoming-animal" as it is a grounded, everyday experience of actual species "making each other up in the flesh," "full of the patterns of their sometimes-joined sometimes-separate heritages."[14] Although domestic cats have developed companion species relations with humans for millennia, they have famously resisted much of the human-structured activity within which horses and dogs seem to thrive. Performance theorist Paul Rae, in an article on the cat in Singapore theatre, takes up this thread when, while adhering to Deleuze and Guattari's perspective of sentimental pet owners,[15] he suggests that Singaporean relations with street-cats might be a special case where the cat's "ungovernable qualities" are "respected." Rae speculates that this might be why the SARS cull

> ... evinced such a strength of feeling, for the affective relation with these cats is one that appeals to an idea of streetwise and sensuous liberty. Fantastical in part, such sentiments nevertheless also hint to an encounter with the non-human whose value in an intensely human-oriented environment is worth fighting for in and for itself.[16]

---

[12] For example, Wolfe cites Gregory Bateson on the ties between human and pre-verbal mammalian communication: "the magnitude of the gesture, the loudness of the voice, the length of the pause, the tension of the muscle and so forth commonly correspond (directly or inversely) to magnitudes in the relationship that is the subject of the discourse." Quoted in Wolfe, *Animal Rites,* 86.

[13] Deleuze and Guattari's (also Spinoza's) term "affect" refers to "[a]n ability to affect and be affected. It is a pre-personal intensity corresponding to the passage from one experiential state of the body to another and implying an augmentation or diminution in that body's capacity to act." Massumi in introduction to Deleuze and Guattari, "Becoming-Intense Becoming-Animal," xvi.

[14] Haraway, *When Species Meet,* 16–17, 25.

[15] Rae, "Cat's Entertainment."

[16] Davis, "Notes for a Singapore Bestiary," 154.

Rae (who prefaces his article by stating that he "is allergic to cats") does suggest a presence of street-cats in Singapore might be worth fighting for in and for itself. But his approach concerns itself with "becoming cat" and "pack" moments in theatre and performance, and does not get into the sniffly business of engaging cats themselves. This was also a path I chose in earlier writing where I explored contested projections of cats in the Singapore body politic, but safely ended with the cat "evading representation ... absent and present, inside and outside, preordained, privileged and at large." I have not had the joy of immersing myself in human-cat contact zones in the ways Hearne and Haraway have with dogs and horses, but from engaging with their explorations it becomes clear that human-street-cat engagements are not a special case, predicated on the demonstrative independence of the cat, but exemplary of numerous, specific "becomings with" between humans and companion species. These "becomings with" can evolve at the same time, even as the species in question is projected through human psycho-social or politico-economic structures in the body politic. "Becomings with" are quotidian co-creations, distinct from the elevated rush accompanying the "becomings-animal" privileged by Deleuze and Guattari.

***

A series of wily, wiry worlds can and do "become with" the same cat in different contexts, and not only "out there" in worlds of "sensuous liberty" conjured between humans and street-cats, but also in the interior jungles of an HDB flat. Daisy believes that many "babies" long for the street, long for the "night breeze"—a phrase she uses to denote the nocturnal zones of street-cats. In the hope that her cats would gradually integrate with street colonies, she installed a cat-flap that opens onto a staircase to the ground floor. But some "babies," instead of returning to the street, took to "loitering" in the open-balcony corridors and staircases of the block and returning to her flat for food and affection—and so joined the ranks of other inside-outside cats with whom they are permitted to "roam."

The newspapers sometimes run stories pertaining to such "in-between" cats— not quite "pets" but not street-cats. They have published letters from residents who do not appreciate the disturbances of such "roamers." Although some town councils turn a blind eye to the "no cats in flats" rule, when incidents go public they react, leading in worst-case scenarios to evictions of both humans and cats from HDB blocks.

Daisy alleges that an elderly man living on the eighth floor of her block has twice captured "roaming babies," and thrown them over the parapet. Once a Malay woman cat-feeder helped bury the cat and clean up the blood. Another time Daisy woke Bangladeshi migrant workers in their quarters beside the block, who helped her dig a grave. Daisy described the perpetrator: "He looks like the devil, an ugly long face with skin stuck to his bones, a dark China man with no double eyelid, the kind that hang[s] out in coffee shops and drive[s] trucks."

Another series of media representations of "cat torturers" may have bled into Daisy's account here, evoking an image of the anti-social man who takes cats and tortures them for pleasure. Groups of cat protectors (again feminised by the media, although men do take part) set up surveillance cameras in areas frequented by known abusers. In a landmark case in 2006, one David Hooi, who repeatedly tortured cats to death, received the maximum 12-month jail sentence after he was captured on camera.[17] Such stories magnify an already gendered prism of fantasies of brutish men who project frustrations with humans in general and women in particular onto cats, and of colonies of cat women who ingeniously hunt them down.

One of the tasks we had that night was to search for a "baby" Daisy had recently released. She was worried this cat had been caught by one of three abusers whom feeders had been tracking in the estate. And so we went outside ... into the "night breeze."

***

Media stereotypes aside, some members of the same constituency targeted by the Graduate Mother Scheme from the 1980s are indeed still single, divorced, gay, childless, or with children moved away. They are now nocturnal cat-feeders as well, touring Singapore streets under the cover of darkness, with bags of Friskies in the boots of their purring Lexusii. In a mirroring of government demographic concerns, such cat-feeders are also zealous advocates of sterilisation, conducting expansive cat-neutering operations.

From their positions of relative influence, these "cat women" were able to corral a surprising amount of support during the SARS episode—even in the government-controlled media, where letters were published to protest the cull.[18] Public memorials were held for culled cats, including one at the five-star Oriental Hotel.[19] Some members of the feminist Association of Women for Action and Research (AWARE)—founded at the time of the Graduate Mother Scheme— also moonlight as "cat women" and attended meetings on strategy vis-à-vis the government and the press.[20]

---

[17]   Kua, "David Hooi Sentenced."

[18]   Although expression of dissent in Singapore is growing, it is still rare—in 2009, Singapore was rated 133rd of 175 countries in the Reporters Without Borders' "Press Freedom Index."

[19]   "Cat Lovers."

[20]   In one of the crisis meetings held during the SARS cull, an AWARE representative was only half-joking when she commented that cats and cat people (like feminists in the past) were now regarded as a "threat to national security." According to the SMS and online rumours, leaders of the animal advocacy groups had been allegedly yelled at and threatened by the head of the government SARS task force for calling a press conference without giving due notice. One SMS I received read, "It was like he feared cats were going to take over!"

More Singaporeans began illegally to take in cats, but once indoors, cats were often caged to stop them from "roaming" or damaging the apartments that were their temporary home. They were maintained until it was thought safe to let them out: cosseted and well-fed, but with their worlds gravely restricted. Luckier individuals were kept in fenced sanctuaries together with dozens of other strays.[21]

This at once coveted and condemned "state of exception"[22] of street-cats during the SARS cull was constitutive of both the government and animal lobby's positions: the government was able to reassert its identity as guardian of public safety while the welfare groups consolidated their position as advocates of compassion and animal liberty. It was also impossible for anyone to capture all of Singapore's street-savvy cats. The still-pervasive public presence of the cat carried extra symbolic weight during this period—as an unwitting target and icon of both parties.

This privileged status of the cat—either at large with a death sentence on its head, captured by the government or lovingly imprisoned in foster care—was further stratified as welfare groups began to differentiate between neutered "community cats" and "ferals." The groups argued that it was often the humanised "community cats" who ended up in traps, while the unneutered, unfriendly, and potentially diseased "ferals" were too wily to be caught. The risk was that once the "community cats" had been trapped, the "ferals" would take over and start to breed.

***

The light was fading as we walked across the street from Daisy's flat. In a group of bushes were members of the first colony on Daisy's rounds. She made a "tsch-tsch-tsch" sound and six skinny but healthy cats, black and tortoiseshell, ran towards her, with tails of various lengths in the air.[23] They surrounded her, making her bend her body down to cat level as they vigorously rubbed and wove around her legs, reaching up on their hind legs to push heads and front shoulders against any body part that was accessible. Individual cats licked and play-bit Daisy's outstretched hands, jumping over each other to enter this haptic zone, lying flat in front of her feet so that she almost fell over them.

---

[21]   Arshad, "Cat Lovers."

[22]   I am referring to Giorgio Agamben's "state of exception," initially developed with regard to the "homo sacer," a person who, according to Roman law, was exiled and a target for execution—to be killed by any citizen but not permitted to be ritually sacrificed. The "homo sacer" inhabited a paradoxical condition of "bare life," at once inside and outside the law. See Agamben, *Homo Sacer: Sovereign Power and Bare Life*, 18, 28, 72. One way of thinking of this exceptional condition of bare life is as an indeterminate, liminal zone, conterminous between man and animal. See Agamben, *Open Man and Animal*, 37–8.

[23]   Southeast Asian cats have a genetic variety of crooked, knotted, and stub tails.

This is the group of street-cats Daisy appears currently closest to and Daisy said this group enacts this frenzied, joyous exchange of fur, scent, saliva, and skin cells every evening. Such "allorubbing" and "allogrooming," according to behaviourists, is how colonies self-become, as each cat contributes to a concoction of scents that each of the colony "wears." [24] *The frequencies and intensities of these rubbings affirm an individual's status; only after the cats had completed this greeting with Daisy did they come to cautiously inspect Jane and me, sniffing our shoes (well-rubbed and scented by our respective cats at home).*

Daisy shares her enjoyment of this human-cat contact zone with another woman (there are six active feeders in the ten blocks of flats of this estate). There were already kibbles and water left out, but Daisy had brought fresh fish. "I make the food myself—it's the same food I eat," she said with pride. Daisy mashed palmfuls of fish between her fingers and crouched down, offering the cats individual portions. They crowded around and ate rapidly, after which they began to groom themselves, Daisy's hands, and each other.

Several people walked past during this elliptical dance of rubbing-eating-grooming. They did not seem surprised to see feeders or cats. Jane explained that the woman in charge of these public areas was sympathetic to the presence of "community cats." Via repeated encouragement and reassurance that feeders in this area were "responsible" and were sterilising cats, Jane had convinced her to not only permit "community cats" to stay, but also turn a blind eye to cats kept in flats. [25]

Daisy said that this group of cats—like the many colonies in her zone—are hidden, sleeping out of the way during daytime. She said that the cats only emerge in areas of human activity towards night, when feeders also appear. We left the cats grooming themselves a little way off the path. As we walked we heard a loud, low, insistent gravelly miaow, and another wiry male, black with a stub tail, ran up alongside us.

"Ah, this is 'Kaypoh cat'!" Daisy exclaimed with some pride, "busybody cat." "He doesn't eat with the rest and follows me each night ... He's not hungry now, look! He just likes to follow!" She offered him a handful of fish but the cat only sniffed it for a few seconds before continuing his low-pitched vocalisations, rubbing around Daisy's legs, running along the path we were to follow, then returning to vocalise and rub again.

As we continued our tour of the estate "Kaypoh" ran sometimes under our feet, sometimes one or two meters ahead or to the side, stub tail high, looking ahead then looking over at Daisy, and filling the space between us with the same, low-pitched

---

[24]   Crowell-Davis, Curtis, and Knowles, "Social Organization in the Cat," 3.

[25]   The ease with which cats, cat-feeders, and other residents coexisted at the time of my visit is emphatically not the case on the rest of the island. That very week we received an urgent SMS from Jurong where private pest control companies were culling "tame, sterilized community cats" after a resident complained. And while fine-tuning this essay, I received a message from Jane about six cats battered to death in Daisy's estate.

vocals. After about 300 metres, Daisy said, "He won't follow much further. He is afraid of a big tom around the next corner." This behaviour—where one cat insists upon a particular, possessive connection and follows a feeder around—is common, as attested by other feeders. While I am not equipped to provide a convincing ethological interpretation, I would venture that the particularly attuned black, stub-tailed, wiry organism of sense and desire that is "Kaypoh" was running the mile and testing the limits of a particular zone of individual interdependence with Daisy.

As Daisy predicted, the sinewy acoustic corridor "Kaypoh" asserted between us ceased at the corner of the block and the young cat crouched, belly to the pavement, watching us as we descended a staircase strewn with piles of kibbles. "That crazy Malay guy, he throws food everywhere and expects cats to eat where everybody is walking!" said Jane.

***

The differentiation between "community" and "feral" street-cats recalls both the gender, race, and class axes of the Graduate Mother Scheme, and demarcations between feeders. Both animal advocacy groups and the government regard with disdain another constituency of cat-feeders who are often elderly and working-class. Although they are not exclusively Malay or Muslim, they are often imagined as such. They are alleged to feed the stray animals "irresponsibly" from messy polystyrene chicken-rice packets as opposed to neat kibbles.

Malays and non-Malays alike pronounce that a Malay-Muslim fondness for felines originates from the Koran, wherein the Prophet Mohammed cut off a corner of his robe rather than disturb a sleeping cat. However, in unofficial, off-the-cuff remarks (never in public statements), elderly working-class Malay Muslim feeders are claimed—in another refraction of state eugenic ideology by younger, middle-class animal advocates—to be "irresponsibly" opposed to sterilisation on religious grounds.

***

We went down the well-lit staircase strewn with kibbles to an open space beside a car workshop, where Daisy believed the "baby" she had recently released was hiding. Three cats—a large ginger male, a slight white female, and a tabby male— emerged, but not the "baby" Daisy was looking for. "I don't think the other feeder takes care when she feeds them. That grey tom is terrorising them." This time there was no intense exchange of rubbing and scents. The cats did not seem afraid, just less interested in Daisy's presence, and initially kept a distance of five or six metres from us.

The slight white cat finally made the first move and jumped onto a raised concrete slab. Now at the level of Daisy's waist, she crouched forward on her front paws while Daisy stroked her. "She prefers it up here as she can see all around and no one can attack her." The cat did seem hungry once she had asserted her place, though she continued to look around every five seconds or so while eating.

I asked Jane whether we should ask Daisy to stop talking so loudly and focus on the cats. Jane replied, "The cats don't seem to mind." She was right. Alongside the noisy dialogue Daisy maintained with Jane and myself, another communication-connection was occurring between Daisy's hands, body, and the white cat she was stroking. The cat similarly appeared to be "multi-tasking," ears moving back and forth, looking up and sniffing the air while keeping close to Daisy's stomach while she ate.

Jane called another feeder on the telephone. Daisy explained how this elderly "Aunty" assisted with kittens and burying dead cats. This seems the case for many feeders who establish contact—online or offline—over cat births, food, sterilisation, shelter, and deaths. This connection over the body of the cat again reflects the domestic care/control preoccupations of the government—something also suggested by the "rebranding" of "street-cats" as "community cats."

***

Alongside efforts to domesticate is an undeniable, vicarious enjoyment of what Paul Rae calls "streetwise and sensuous liberty" or Daisy calls the "night breeze," as online pleas request help to relocate cats to places where they can "live freely once more." Respect for the freedom of the cat to live as a cat is also recognised, albeit somewhat ironically, in the term for the street-cat sterilisation programme "Trap-Neuter-Release." Haraway writes of the integration of such romanticised "free spirits" as feral cats into human technocultures, from the "alien abductions" and invasive body surgery accompanying sterilisation, to the industrially processed kibbles the cats consume.[26] There is a related irony in the way these extensive, even excessively organised networks of Singapore feeders harness the disciplinary discourses and technologies of the state, in order to conjure and police space-and-time zones "out there in the night breeze" in which street-cats can be free.[27]

The SARS cat-culling incident then, recasts the nexus of gender, race, and class at stake in the Graduate Mother Scheme, but this time in and around the exceptional status and infectious appeal of the street-cat. Indeed it would be hard to find a creature more antithetical to government ideology than the street-cat of

---

[26] Haraway, *When Species Meet*, 277–8.

[27] HDB public spaces are not just the purview of cats and cat-feeders. They are sites for all public-private activities which used to take place on *kampung* (village) verandahs and yards. The "void deck" spaces on the ground level of HDB blocks host ball games, music groups, weddings, and funerals. Hindus and Taoists occupy the base of trees next to blocks, often right next to cat-feeder stations. Human-cat-feeder zones are therefore overlaid with gods, spirits, and the energy of everything else that *unlike* the cat is not able to infiltrate the modern order of the HDB flat.

popular imagination. It does not follow the grid system of the city. It does not walk on a leash and cannot be trained. It is unpredictable, nocturnal, transgressive. It has sex very loudly and uses its sexuality "irresponsibly," reproducing out of control.

A final gender/race/class dimension to the crisis was the way that migrant workers of minority ethnic groups were brought in to carry out the cull. Bangladeshi migrants were allegedly paid per captured cat. Stories circulated via that "feral" medium of SMS of missing community cats being crammed into dumpsters—stories that were accompanied by a brutalisation of the migrant, as not being civilised enough to cull animals humanely. Yet these are the same migrant labourers who are transported like livestock in trucks and leaking ships around Southeast Asia, inhabiting many of the same "out there" spaces as the street-cat, stray dogs, and other "vermin."[28]

Figure 13.1 from *The Straits Times* reveals a different gender/race/cat dynamic, depicting Singaporean minority male Malay-Muslim "pest control officers" who have been asked to catch the deviant cat—even as Malays generally hold cats in high regard. Equivalences between cats, class, mice, and men are evoked as the pest control employees crawl under a Mercedes-Benz—the car of choice of the Chinese nouveau riche.

I am not simply arguing that the SARS cat-culling incident was a repeat performance of the 1980s Graduate Mother Scheme, or a time when equivalences were drawn between states of exception for Singapore street-cats and other marginalised beings. The contagion[29] unleashed by the movement to save street-cats was fraught with complexities—both progressive and reactionary. What is more, attempts are still being made to claim the narrative of this crisis, with all sorts of negative and positive projections jumping race/gender/class/species barriers and being policed back into place.

Invocations of the cat in the Singapore body politic took on another affective register in 2009 when thousands of women (and not a few male supporters) attended an Extraordinary General Assembly (EOGM) called by the feminist organisation AWARE. The EOGM came about after an infiltration and takeover of this Non-Government Organization (NGO) by Christian fundamentalists opposed to AWARE's public and progressive stance on homosexuality, abortion, and sex education.

Progressive Singaporeans rallied behind the "old guard" who had previously led and shaped the organisation, and similar to the SARS cat-culling incident, the press was surprisingly supportive of this group.[30] However a few senior, male

---

[28]   For the spatial-temporal overlay of animals and migrant human worlds, see Philo and Wilbert, *Animal Spaces, Beastly Places.*

[29]   "Contagion" provokes mutant, impure alliances: "[A]nimals are packs, packs form, develop and are transformed by contagion." Deleuze and Guattari, "Becoming-Intense Becoming-Animal," 242.

[30]   The government has since issued statements which, if not in support of the feminist or the gay lobby (homosexuality is against Singapore law), warned of the expansionist practices of new churches. "Transcript of National Day Rally 2009 Speech."

Fig. 13.1   "It's 1-0 to the cat on the left as pest control officers Hairudin
                Mohammad Saad and Azmi Mohamad miss a fleeing stray in Yishun.
                But its compatriot is not so lucky in the cat and mouse game." *The
                Straits Times*, May 2003. Photograph by Malcom Mcleod. Image
                reproduced courtesy of Singapore Press Holdings.

commentators sought to dismiss the struggle as a "cat fight." As the leader of the
fundamentalist camp was named Josie Lau, the infiltrators were termed "Josie and
the Pussycats" in blogs and satirical commentaries. [31] And during the tumultuous
events of the EOGM itself, where putsch leaders attempted to retain control while
being heckled by thousands of "old guard" and gay rights supporters chanting,
"We Are AWARE!" a male church member took the mike: "Ladies! Please! This
is not how Singapore ladies should behave! We need to be civilised! We have
to discuss these things rationally! I received a shock outside when one of your
supporters ... Do you know what she did? She turned round and hissed at me! Just
like a cat!"

***

*We walked on without finding the "baby" Daisy was looking for. She said, "If
there is a cat I haven't seen I can't sleep. I think I can hear them and have to go
back to look."*

*We climbed to the top of a small hill and Daisy exclaimed, "The cats up here
are big and fat with doll eyes—very cute!" Indeed this colony did resemble poster
cats for a "community cat" campaign; fat, sleek, lying in an open public space (in
HDB-speak the "void deck") on the ground level of a housing block where a group
of elderly residents was sitting and chatting. Daisy went to greet the residents. A
girl in school uniform walked past and made the "tsch-tsch-tsch" sound Daisy had
made earlier. Two cats raised their heads and sniffed the air around the schoolgirl,
then lay down again. Another cat rolled over and back again.*

---

[31]   *Josie and the Pussycats* is a popular cartoon TV series from the 1970s, featuring an
all-girl pop band in a variety of light-hearted adventures.

That's the
cats taken
care of...

**The old
timers
are next!**

I received a rather official letter   were mugging unsus-   and defecate and uri-
this week. So, I thought it was my   pecting rats and steal-   nate in public places.
civic duty to reproduce it here ...   ing their fish scraps.   Of course, these

Fig. 13.2   In a somewhat different take, equivalences between cats and
marginalised humans were not overlooked. *TODAY*, 7–8 June 2003.
Courtesy of MediaCorp Press.

*A young Indian neighbour walked past smiling. "I haven't seen these cats for
days, then you come and they are all here!" Daisy beamed. Actually, this was the
first place where cats seemed to have already established themselves before Daisy
arrived. Again they didn't seem hungry. Five or six individuals sniffed at Daisy's
bags. But their dance was a slow one, circling around before finally moving to rub
her legs. A luscious ginger cat groomed his chest. Another pair of grey and white
cats with symmetrical markings did not "greet" us but rubbed up against objects:
concrete pillars and stools in our proximity.*

*"I just love these succulent fatty bom-boms!" Daisy cried, as she energetically
ruffled the back of a large male with a long tail. I remarked that many of these
"community cats" had long, straight tails. Jane replied that there was probably
some artificial selection going on where "cuter" cats were given more food and
attention. Other feeders have made similar anecdotal observations that there are
far fewer cats with stub or crooked tails on Singapore streets than there used to be.*

*Daisy disappeared and came back carrying a thin cat with features similar to
the others. "This cat gets bullied by the rest and doesn't get food." Like "Kaypoh,"
this cat was not fully integrated into the colony but seemed to have a particular*

connection with Daisy. The cat at first shrank away from Daisy's hand, then slowly
began to respond to her stroking, rubbing against a nearby wall but not venturing
into the space around. For the first time in the evening Daisy lowered her voice as
she spoke softly to this cat. The cat ate quickly while she was close, but darted off
into the dark when another cat approached.

***

In a series of conversations I had with Paul Rae, he ventured that both the cat-
feeder's drive to connect with street-cats at night, and such manifestations of excess
in the body politic as the AWARE EOGM and the SARS cat-culling incident,
are all related symptoms of state rationality. He argues that Singapore's regulated
domesticity becomes so stifling that eruptions of unformed affect or what he calls
"hysteria" are inevitable.

I agree with the outline of Rae's hypothesis (if not his insistence on the
historically-gendered term "hysteria"). However, I am not convinced by an analysis
where All is either "domestication" or unformed (and thereby un-narratable)
affect; where All is either rationality or "hysteria." Such an analysis seems a too
easy inversion of the Singapore government's own self-legitimating rhetoric of
comfort-through-control or chaos.

To return to the AWARE EOGM: what I observed, without denying the
inescapable cathartic rush of what Rae might call the "pack," were multiple,
but very specific, refractions of the emancipatory possibilities of this energy.
Alongside the boos, jeers, whistles, and chanting of "We are AWARE!" and the
camp gesture of the woman who so startled the gentleman from the church with
her "hiss" were women who spoke soberly into the microphone about growing
up gay in fundamentalist Christian families. Popular actress Irene Ang spouted
bawdy jokes on sexuality and Christian prudery. Another woman pronounced
this battle to be less about religion versus homosexuality, than about corporate
versus NGO culture. And there was an emphatic response to the same male church
member, when an AWARE member marked out a strategic space for more emotive
"feminised" communication: "We are not as you imply, sir, out of control. But
in this context it is your rules of rational debate which are not appropriate. The
reason why they are not appropriate is because we are passionate! … This is what
it sounds like when people are passionate!"

In order to elaborate on the range of positions being staked within what Rae
outlines as yet another moment of "hysteria," I have slipped away in the end of
this chapter from invocations of or communions with the body of the cat. But my
point here is this: it seems as much a disfavour to sweep the AWARE EGOM,
the SARS cat-culling incident, and the individuated, micro-entanglements of
humans and street-cats together as pressure-valve or "package" responses to state
domestication, as it would be to dismiss Daisy's relationships to her "babies" as
simply Oedipal and reactionary.

My challenge in writing this chapter has been to remain both loyal to the many specific registers of "becomings with" of feeders and cats, and to elaborate on a contested rash of possibility arising from such larger public moments as SARS, without deadening the potential of either with a too-definitive analysis. I hope the above has revealed some of the transgressive transmissions of political energy and projection that the SARS cat-culling incident unleashed in the body politic, while tracing some of the myriad co-creations of haptic, olfactory, and acoustic worlds by humans and cats as they thread through Singapore's public and private spaces—from the jungled interiors of HDB flats to the "night breeze" of "Kaypoh" and the "succulent fatty bom-boms."

Chapter 14
When Is Nature Not?

Tim Low

Domination of the world by humans is usually thought of as a unique process in the history of life on earth. But there is fossil evidence to show that something similar occurred about six to eight million years ago. It was the dramatic rise to dominance in tropical regions of a group of fast-growing grasses, leading to the creation of savanna habitats.

In what was geologically a wink in time, during the Miocene, savanna grasses took over half the land on earth once occupied by forest.[1] Changed forever was the face of Africa, of Asia, of North America, and of Australia.[2] Biologists have described "a dramatic near-synchronous expansion of C4 grasses around the world within about a million years."[3] Their mention of "C4" refers to the highly efficient photosynthetic pathway of these grasses.

Human domination of the world has resulted in many animal extinctions, and so it was with these grasses. Various forest-dwelling mammals disappeared from Africa, Europe, Asia, and North and South America, apparently because of habitat loss.[4] For many leaf-eating mammals that survived, a change in diet to more grass can be detected in carbon isotopes in the enamel of their fossil teeth.

Human domination was made possible by many tools adopted by humans, and I would argue that savanna grasses also relied on a "tool"—fire. Fire was not new at this time. Fossil charcoal in large amounts can be traced back more than three hundred million years.[5] But fires increased dramatically in the late Miocene as grasses spread. Charcoal layers in the sea point to vast fires in Asia and Africa, and ash in the Atlantic and Pacific can be identified as coming from grasses.[6]

An increasingly monsoonal climate is thought to have aided grasses by increasing the opportunities for fires.[7] Monsoons feature a pronounced wet season, ideal for growth, followed by a dry season during which fires lit by lightning can spread over large areas. Fires in many savannas burn each year.

Grasses are often the plants that benefit most from fire. Fire is a major force by which dominant grasses compete with shrubs and trees because they recover

---

[1]   Bond, Woodward, and Midgely, "Global Distribution," 525.

[2]   Some Australian ecologists dispute that fire maintains Australia's savanna habitats, so I have listed Australia last.

[3]   Beerling and Osborne, "Origin of Savanna Biome," 2023.

[4]   Damuth and Theodor, "Miocene Ungulates."

[5]   Scott, "Pre-Quaternary History."

[6]   Beerling and Osborne, "Origin of Savanna Biome."

[7]   Osborne, "Atmosphere, Ecology and Evolution."

more quickly from fires than woody plants. In grasses you have the evolution of flammability: they have evolved to burn. Their short-lived leaves accumulate as fuel beneath living plants. In what is a form of symbiosis, grasses promote fire and fire promotes grasses.

American ecologist Robert Mutch was the first to advance the notion of flammability as an adaptation favoured by natural selection, in 1970.[8] His theory has won support in many subsequent papers, with titles such as "Kill Thy Neighbour" and "Are Some Plants Born to Burn?" The evolution of flammability is very evident in Australia, where fires allow spinifex to displace mulga woodlands, buttongrass to displace shrubs, and black spear grass to displace softer grasses. Buttongrass in particular seems made to burn—it will ignite at temperatures below 1°C, at night during fog or when frost is forming.[9]

Late Miocene grasses set the world on fire, and planet earth has been burning ever since. Fires lit by lightning, and more recently by people, have benefited grass-exploiting animals, including humans, as well as the grasses themselves. Fire was the most powerful tool at the disposal of early humans. *Homo sapiens* began by burning forests and grass, and now burns coal, oil, and gas. Without fire, and without plant-based fuels to burn, humans would never have achieved such dominance. The parallels between grass success and human success are thus substantial.

One could argue there is one big difference in that humans are one species and grasses many. But the rise of humans has not been the rise of a single species. Our crops, livestock, and companion animals have multiplied along with us. And our success has been exploited by many others, including rats, mice, cockroaches, lice, crows, seagulls, vast numbers of weeds, and some exceptionally nasty ants.[10]

Our rise could even be classified as a second stage in a single process, because we would not have gone anywhere without grasses. We evolved in savanna. Most of our staple foods—wheat, rice, corn, and sugar—come from grasses, and grasses feed our livestock. But it must be said that our most important grasses were not part of that Miocene grass explosion.

Humans have a powerful relationship with fire, just as we do with grasses. Indigenous people often lit fires. Most of the early explorers in Australia saw evidence of deliberate fire.[11] Fire was used by people all around the world.

In 1969 Rhys Jones published his milestone paper in which he claimed that Aborigines had farmed the land by use of fire.[12] They were fire-stick farmers. "Aborigines as farmers" was a wonderful way to raise their status in mainstream Australian eyes. European settlers had justified their land grabs by asserting that

---

8   Mutch, "Wildland Fires and Ecosystems."
9   Low, "Born to Burn," 24.
10   Low, *New Nature*.
11   Pyne, *Burning Bush*, 121.
12   Jones, "Fire-Stick Farming."

the Aborigines had failed to manage or improve the land in any way. Rhys Jones was refuting that argument.

But now the concept of Aboriginal management has become so entrenched it has led to the idea of Australian vegetation as a human artefact. Tim Flannery pushed this in a 2003 essay: "it's not an exaggeration to say that Aboriginal fire and hunting literally made the Australian environment that the Europeans first encountered. It was a vast, 47,000-year-old human artefact, designed to provide maximal food and comfort to its inhabitants in the most sustainable manner."[13]

But Flannery overstates the importance of ignition and ignores the role of flammable vegetation in making fires easy to light. The Aborigines could use this method only because flammable vegetation had evolved millions of years before their time. Botanist David Bowman has put it like this: "Aboriginal people played an important part in the making of flammable Australia as we know it, but they did not trigger the relentless fire cycle; rather, over some 40–70 [thousand years] they learnt to harness the naturally-occurring fires to their economic advantage."[14]

Indigenous Australians extended a natural process by increasing the frequency and reducing the reach of fires, but they did not initiate something new. They appeared to wield enormous power only because Australia carries so many flammable plants. Among these are eucalypts, paperbarks, and many other plants with highly flammable bark and flammable oil in their leaves. Flammability helps eucalypts outcompete rainforest trees, cypress pines, and she-oaks, and it also helps grasses compete with eucalypts.

Instead of talking about "Aboriginal fires" we should acknowledge the mutual relationship that existed between people and particular plants. Humans needed the grasses to light fires, but grasses didn't need humans to the same extent because there is lightning as an alternative source of ignition.[15] To call the Australian bush an "artefact" in this situation is to misunderstand completely the relationship between those who were doing the burning and those who were being burnt.

The *Oxford English Dictionary* defines an "artefact" as "the product of human art and workmanship ... as distinct from a similar object naturally produced." This is not a valid way to think about Australian landscapes in which human ignition was only one force among many. Environmental philosopher Val Plumwood has strongly attacked Flannery's image of the Australian vegetation as something created: "The picture of Australia as a human product presents creativity as the prerogative of the human and denies the role of forces much older and more powerful than the human in shaping the continent." To quote Plumwood again: "Counting something (e.g., a place) as purely human (or 'cultural') when it

---

[13]   Flannery, *Beautiful Lies*, 41.

[14]   Bowman, "Australian Landscape Burning," 261.

[15]   The world is hit by eight million lightning strikes each day. Scott, "Pre-Quaternary History of Fire," 284.

involves the labor of nature jointly with human labor hides or denies the work of ecological systems and human dependency relations on it."[16]

This "artefact" example, and its popular reception, shows how inadequate humans can be at thinking about nature as an active agent. Other examples abound. *The End of Nature*, an important best-selling book, provides an extreme example.[17] Its author, Bill McKibben, argues that because human-induced climate change has impacted upon nature everywhere, nature has ceased to exist. McKibben sought to jolt people's concerns about climate change, but his conclusion, that nature ceases to exist as soon as there is some human influence, is nonsensical. It denies the reality that humans once lived in most habitats. McKibben really means "wilderness" when he says "nature," but this word is also problematical.[18] Does humanity cease to exist if it is influenced by nature? Why should the reverse be true?

Another heavy-handed perspective comes from Michael Pollan, in his book *The Botany of Desire*: "What if those potatoes and tulips have evolved to gratify certain human desires so that humans will help them to multiply? What if, in other words, these plants are using us just as we are using them?"[19] This is like arguing that grasses exploit people when fires are lit; it simply shifts causality from people to plants, while perpetuating the simplification.

In these three examples, which have come from prominent, influential writers, the crudeness of the thinking suggests a widespread blind spot, a failure to properly consider nonhuman causality. To paraphrase the three examples:

1. Aborigines lit fires, therefore Australian vegetation is an artefact;
2. Humans have changed the climate, thus nature has disappeared;
3. Crops benefit from humans, therefore crops rule humans.

A more subtle example of fudged thinking runs through our thinking about extinction. We are in the midst of a global extinction event, during which more mammals have gone extinct in Australia than anywhere else.[20] Humans are clearly to blame. But Australians are also told that foxes and cats caused the extinctions. It so happens that some of the vanished animals were never seen by Europeans, much less killed by them. Some extinctions took place in deserts where no axe or hoof ever went. Parallel narratives exist: foxes and cats did it; people did it.

Weeds are another environmental problem where agency can be contested. A weed expert[21] working on camphor laurels told me that all the hundreds of thousands of feral camphor laurels in the Bellingen valley in New South Wales owe descent from six trees planted a century ago at the local school. Native birds

---

16   Plumwood, "Concept of Cultural Landscape," 132–8.

17   McKibben, *End of Nature*, 43

18   Low, *New Nature*, 37

19   Pollan, *Botany of Desire*, back cover.

20   Johnson, *Australia's Mammal Extinctions*.

21   Judy Davies, personal communication.

feeding on the school ground berries have spread the seeds far and wide. Many of the world's worst weeds happen to have seeds that are widely spread by birds.

If it is acceptable to talk about Aboriginal people farming the land with fire, surely one may talk about birds sowing camphor laurel seeds. Birds get the habitat they want when they drop seeds, in what represents a mutually beneficial relationship with plants. With six trees planted by people, and hundreds of thousands planted by birds, who is more accountable for the weed problem in Bellingen Valley? One could ask the same question about the weed problems caused by lantana, blackberries, prickly pear, and other major weeds.

Humans are held responsible for weed problems, and for extinctions, because human actions—the introduction of foxes and foreign plants—came first, leading to subsequent problems. But if you adopt that line of argument, questions must be raised about whether humans are responsible for the fires they light, because the evolution of flammability preceded the lighting of fires. Rainforest had declined because of grass and eucalypt-fuelled fires long before people reached Australia; indigenous Australians accelerated the process but were not the main cause of rainforest loss.[22]

I believe we should accept responsibility for extinctions, and for weeds, but only because we have comprehension, foresight, and morality to guide us, not because we are entirely to blame. We are the only participants who can intervene and for that reason we should. It has been unlucky for most life forms on earth that a species as powerful as us has arrived, but fortunate for some that we pay some heed to moral values. If killer ants had risen to world dominance they would not have set aside safe areas for other species, nor devised laws to protect them.

To think more realistically about the world we should acknowledge the power of nonhuman agency. In my book *The New Nature* I give many examples of animals as "ecosystem engineers" shaping landscapes, including elephants, seals, wombats, beavers, seabirds, termites, and corals.[23] But most people, including experts, are so reluctant to recognise nonhuman influence that animal agency is regularly attributed to people. In the Northern Territory the salinisation of coastal wetlands has been blamed on human-induced sea level rise, when the main cause has been water buffaloes creating swim channels that let in the sea,[24] and in north Queensland, debates have raged about whether certain mounds of shells represent indigenous campsites or old nest sites of orange-footed scrubfowl.[25]

Our language limits our thinking about nonhuman agency. On small islands where seabirds breed, nutrient levels can rise so high that native plants die and soil erodes into the sea.[26] We should be able to talk about seabirds "polluting" island

---

22   Bowman, *Australian Rainforests*, 285.

23   Low, *New Nature*, 49.

24   Low, *Climate Change and Invasive Species*, 15.

25   Bailey, "Hens' Eggs and Cockle Shells."

26   Low, *New Nature*, 53.

soil, but the word "pollution" is almost never applied to nonhuman wastes, and "erosion" is seldom applied to animal impacts.

Christmas Island in the Indian Ocean is an example of a place where animal agency is especially obvious. The island's red land crabs are such efficient consumers of seeds and seedlings that they produce an open rainforest structure.[27] But invasive yellow crazy ants reached the island long ago on ships, and they now prey heavily on the crabs and have eliminated them from some areas. Where ants dominate, the rainforest is now much denser with seedlings and saplings.[28] The difference between crab-ruled and ant-ruled rainforest is obvious—one is easy to walk through and the other too thick. Christmas Island must have been a much better place for plants and insects before it was colonised by crabs. Not only do the crabs consume seeds and seedlings, they also eat fallen leaves that would otherwise shelter small invertebrates. Red crabs have much in common with humans, as abundant domineering animals that alter whole landscapes when they divert resources to their own ends.

I do not think humans are as different from other species as we are usually thought to be, because I see so many parallels between us and crabs and ants and grasses, and because I can see that we often exaggerate human impacts and underestimate other players.[29]

A couple of years ago I wrote a report for the Australian government about climate change and invasive species.[30] Of all the introduced plants and animals I considered, the species that concerned me the most was gamba grass, a giant African grass imported into northern Australia for pasture. Farmers want this grass because it produces far bigger leaves than any native grass, but in northern Australia, what is intended as food for cows often becomes fuel for very big fires. Gamba grass fires are so hot that eucalypt trees are often killed.[31] Gamba is only one of many introduced grasses that can shift fire regimes into a self-perpetuating alternative stable state.[32] The Queensland government assessed the risk it poses and reached an alarming conclusion: "If large areas of northern Australia become dominated by gamba grass, the associated fire regime is predicted to transform Australia's eucalypt-dominated tropical woodlands into tree-free grasslands."[33] Climate change, by increasing the fire risk, makes that more likely. Here is a plant that poses as much threat to biodiversity as a thousand bulldozers. Governments have since moved to ban it, although existing plantings remain (and are spreading).

This example shows that when planning for climate change we need to consider nonhuman agency. It is not enough just to ask: "How will climate change directly

---

27   O'Dowd, Green, and Lake, "Invasional Meltdown."
28   Ibid.
29   Low, *New Nature*, 49.
30   Low, *Climate Change and Invasive Species*.
31   Rossiter et al., "Testing Grass-Fire Cycle."
32   D'Antonio and Vitousek, "Biological Invasions."
33   Csurhes, *Assessment of Potential*, n.p.

affect humans and other species?" We should also ask how climate change could benefit species that cause harm. In some ecosystems the main threats to biodiversity may well come, not from climate change directly, but from undesirable species that thrive under climate change.

It is therefore important to have a suitable framework for thinking about the future. We should see ourselves as not always operating alone when we cause environmental harm. All too often, perhaps most of the time, we have partners in crime—species that exploit situations and worsen outcomes. The number of animal and plant "pests" contributing to conservation problems is vast, and includes rats, mongooses, elephants, monkeys, deer, pigs, birds, snakes, centipedes, starfish, bees, vines, and various pathogens.[34]

Here are some suggestions for better thinking about this topic. We should stop always thinking about nature in the singular, and more often as a "they." When a farmer razes a forest to create a pasture, it is not true that all birds suffer. The pastures so created suit birds such as magpies that dislike forests. What this shows is that there is no one response to us—or to anything else—by other species, but rather many individual responses, because each species has its own needs. We should think of nature as a multiplicity rather than as a single thing.

We should beware the simplistic duality of humans versus nature. The concept of harmful humans and innocent nature is a useful rhetorical device for advancing nonhuman interests, but ultimately it limits our thinking. We have to be able to recognise when nonhuman species behave unhelpfully (and when humans behave well). We have a moral obligation to prevent such awful outcomes as rats on islands exterminating birds by eating their chicks.

We need new ways of thinking about causality, which recognise how often it is shared. Humans are implicated in recent extinctions but they are not the only villains. Nor are they the only reason why fires burn so far or weeds spread.

Finally, as I keep emphasising, we should acknowledge the active agency of nature. Many forces are at play around us but we tend not recognise them. For as Val Plumwood says: "Many philosophers try to impose consciousness as a condition of agency."[35] But unconscious agency can be extremely powerful, as Freud concluded. Perhaps that helps explain our reluctance to think more carefully about these issues. Our conscious minds prefer a simple picture of a consciously ruled world. But ultimately, a world in which animals and plants are active agents of change is more satisfying.

---

[34] Low, *Feral Future*.

[35] Plumwood, "Concept of Cultural Landscape," 124.

Bibliography

Acres, Avis. *Opo the Gay Dolphin*. Auckland: Reed, 1956.

Adams, Carol J. *Neither Man nor Beast: Feminism and the Defense of Animals*. New York: Continuum, 1995.

———. *The Sexual Politics of Meat: A Feminist Vegetarian Critical Theory*. New York: Continuum, 1990.

Adams, Carol J., and Josephine Donovan, eds. *Animals and Women*. Durham: Duke University Press, 1995.

Adams, Brooks. "The Coldest Cut: Sue Coe's Porkopolis." *Art in America* 78, no. 1, (1990): 126–9.

Agamben, Giorgio. *Homo Sacer: Sovereign Power and Bare Life*. Translated by Daniel Heller-Roazen. Stanford: Stanford University Press, 1998.

———. *The Open Man and Animal*. Translated by Kevin Attell. Stanford: Stanford University Press, 2004.

Aloi, Giovanni. "Angela Singer: Animal Rights and Wrongs." *Antennae: The Journal of Nature in Visual Culture* 7 (2008): 10–17. http://www.antennae.org. uk/ANTENNAE%20ISSUE%207.doc.pdf (accessed 28 September 2009).

Alpers, Antony. *A Book of Dolphins*. London: J. Murray, 1960.

American Library Association. "'And Tango Makes Three' Tops ALA's 2006 List of Most Challenged Books." American Library Association, 6 March 2007. http:// www.ala.org/Template.cfm?Section=news&template=/ContentManagement/ ContentDisplay.cfm&ContentID=151926 (accessed 25 May 2009).

American Pet Products Manufacturers Association. "National Pet Owners Survey: The 2005–2006 APPMA National Pet Owners Survey Cites Largest Growth in Pet Ownership in Two Decades." *APPMA Advisor*, May 2005. http:// www.americanpetproducts.org/newsletter/may2005/npos.html (accessed 30 November 2009).

American Red Cross. "Pets." http://www.redcross.org/portal/site/en/menuitem. d8aaecf214c576bf971e4cfe43181aa0/?vgnextoid=0879a06d7565b110Vgn VCM10000089f0870aRCRD&vgnextfmt=default (accessed 5 November 2009).

Anderson, Kay. "A Walk on the Wild Side: A Critical Geography of Domestication." *Progress in Human Geography* 21, no. 4 (1997): 463–85.

Anon. "Sue Coe Interview." *3 x 3: The Magazine of Contemporary Illustration* 5 (n.d.). http://www.3x3mag.com/sue_coe.html (accessed September 28, 2009).

Animal Studies Group, The, ed. *Killing Animals*. Urbana: University of Illinois Press, 2006.

Archer, John. *Ethology and Human Development*. Hemel Hempstead: Harvester Wheatsheaf, 1992.

———. "Why do People Love their Pets?" *Evolution and Human Behaviour* 18 (1997): 237–59.

Armbruster, Karla. "Creating the World We Must Save: The Paradox of Television Nature Documentaries." In *Writing The Environment: Ecocriticism and Literature*, edited by Richard Kerridge and Neil Sammells, 218–38. London: Zed, 1998.

Armstrong, Philip. *What Animals Mean in the Fiction of Modernity*. London: Routledge, 2008.

———, and Laurence Simmons, eds. *Knowing Animals*. Leiden: Brill, 2007.

Arshad, Arlina. "Cat Lovers Save 2,000 Strays from Culling." *The Straits Times*, 8 June 2003.

Associated Press. "House Passes Pet Evacuation Bill." *CBS News*, 22 May 2006. http://www.cbsnews.com/stories/2006/05/22/politics/main1644260.shtml (accessed 5 November 2009).

Australian Broadcasting Commission. http://www.abc.net.au/melbourne/stories/ s1569606.htm (accessed 1 May 2007; page now discontinued).

"AVA admits: Yes We are Culling Stray Cats." *The Straits Times*, 24 May 2003.

Aviram, Amitai. "The Placebo Effect of Law: Law's Role in Manipulating Perceptions." *George Washington Law Review* 75 (2006): 54–104.

Babe. Directed by Chris Noonan. Millers Point, NSW: Universal Pictures Australia, 1995.

Bailey, G. "Hens' Eggs and Cockle Shells: Weipa Shell Mounds Reconsidered." *Archaeology in Oceania* 26 (1991): 21–23.

Baker, Steve. *Picturing the Beast: Animals, Identity and Representation*. Manchester: Manchester University Press, 1993.

———. *Picturing the Beast: Animals, Identity and Representation*. Foreword by Carol J. Adams. Urbana: University of Illinois Press, 2001.

———. *The Postmodern Animal*. London: Reaktion, 2000.

———. "'You Kill Things to Look at Them': Animal Death in Contemporary Art." In Animal Studies Group, *Killing Animals*, 69–98.

Balaskó, Marta, and Michel Cabanac. "Behavior of Juvenile Lizards (*Iguana iguana*) in a Conflict Between Temperature Regulation and Palatable Food." *Brain, Behavior and Evolution* 52 (1998): 257–62.

Balcombe, Jonathan P. "Laboratory Environments and Rodents' Behavioural Needs: A Review." *Laboratory Animals* 40 (2006): 217–35.

———. *Pleasurable Kingdom: Animals and the Nature of Feeling Good*. London: Macmillan, 2006.

Balcombe, Jonathan P., Neal Barnard, and Chad Sandusky. "Laboratory Routines Cause Animal Stress." *Contemporary Topics in Laboratory Animal Science* 43 (2004): 42–51.

Balluch, Martin. "Animals Have a Right to Life." *Altex* 23 (2006): 281–6.

Bamford, Helen. "'Canned Hunting' Industry Flourishing Despite Outcry." *Weekend Argus*, 30 October 2004.

Barr, Stuart, Peter R. Laming, Jaimie T. A. Dick, and Robert W. Elwood. "Nociception or Pain in a Decapod Crustacean?" *Animal Behaviour* 75 (2008): 745–51.

Bate, Jonathan. *Romantic Ecology: Wordsworth and the Environmental Tradition.* New York: Routledge, 1991.

Becoming Animal: Contemporary Art in the Animal Kingdom. Edited by Nato Thompson. Cambridge: MIT Press, 2005. An exhibition catalogue.

Beerling, D. J., and C. P. Osborne. "The Origin of the Savanna Biome." *Global Change Biology* 12, no. 11 (2006): 2023–31.

Bekoff, M. "Aquatic Animals, Cognitive Ethology, and Ethics: Questions About Sentience and Other Troubling Issues that Lurk in Turbid Water." *Diseases of Aquatic Organisms* 75 (2007): 87–98.

———. "Wild Justice and Fair Play: Cooperation, Forgiveness and Morality in Animals." *Biology and Philosophy* 19 (2004): 489–520.

Bentham, Jeremy. *An Introduction to the Principles of Morals and Legislation.* London: T. Payne, 1789.

Bernard, Claude. *An Introduction to the Study of Experimental Medicine.* Translated by Henry Copley Green. Introduction by Lawrence J. Henderson. New York: Dover Publications, 1957.

Berridge, Kent C., and Morten L. Kringelbach. "Affective Neuroscience of Pleasure: Reward in Humans and Animals." *Psychopharmacology (Berlin)* 199 (2008): 457–80.

Bhabha, Homi K. "Foreword." In *The Wretched of the Earth* by Frantz Fanon, translated by R. Philcox, vii–xii. New York: Grove Press, 2004.

Blanchard, Robert J. "Animal Welfare Beyond the Cage … And Beyond the Evidence?" *Journal of Applied Animal Welfare Science* 13, no 1 (2010): 89–95.

Blum, Scott, and Roxane Cohen Silver. "Why is it Important to Allow People to Evacuate Disaster Areas With Their Pets?" American Psychological Association Online Public Policy Office: Psychological Research on Disaster Response. http://www.apa.org/ppo/issues/katrinaresearch.html (accessed 5 November 2009).

"Body of Baboon Killed by Banned Pesticide is Exhumed, Incinerated." *Mercury*, 28 August 2006.

Boice, Robert. "Burrows of Wild and Albino Rats: Effects of Domestication, Outdoor Raising, Age, Experience, and Maternal State." *Journal of Comparative and Physiological Psychology* 91 (1977): 649–61.

Bond, W. J., F. I. Woodward, and G. F. Midgely. "The Global Distribution of Ecosystems in a World without Fire." *New Phytologist* 165 (2005): 525–38.

Bowman, David. "Australian Landscape Burning: A Continental and Evolutionary Perspective." In *Fire in Ecosystems of South-West Western Australia*, edited by I. Abbott and N. Burrows, 107–18. Leiden: Backhuys Publishers, 2003.

———. *Australian Rainforests: Islands of Green in a Land of Fire.* Melbourne: Cambridge University Press, 2000.

Bradshaw, G. A. *Elephants on the Edge: What Animals Teach Us about Humanity.* New Haven, Connecticut: Yale University Press, 2009.

Bradshaw, Peter. Review of *Happy Feet* (Warner Bros movie). *Guardian*, 8 December 2006, http://www.guardian.co.uk/culture/2006/dec/08/2 (accessed 25 May 2009).

Breytenbach, Karen. "SPCA to Investigate Killing of Relocated Tokai Baboon by Pack of Dogs at Dutoitskloof." *Cape Times*, 12 June 2007.

Bridges, John Henry. "Harvey and Vivisection." *The Fortnightly Review* 26 (1876): 1–17.

Browning, Gina. "Pet Evacuation Act Passed by House; Senators Urged to Cosponsor and Support Act." SPCA: Serving Erie County, NY, May 25, 2006. http://www.yourspca.org/site/News2?page=NewsArticle&id=5727 (accessed 5 November 2009).

Brüne-Cohrs, Martin, Ute Brüne-Cohrs, and William C. McGrew. "Psychiatric Treatment for Great Apes?" *Science* 306 (2004): 2039.

Bryld, Mette, and Nina Lykke. *Cosmodolphins: Feminist Cultural Studies of Technology, Animals, and the Sacred*. London: Zed Books, 2000.

Bshary, Redouan, and Alexandra S. Grutter. "Punishment and Partner Switching Cause Cooperative Behaviour in a Cleaning Mutualism." *Biology Letters* 1 (2005): 396–99.

Bshary, Redouan, and Daniel Shäffer. "Choosy Reef Fish Select Cleaner Fish That Provide High-Quality Service." *Animal Behaviour* 63 (2002): 557–64.

Bshary, Redouan, and Manuela Würth. "Cleaner Fish *Labroides dimidiatus* Manipulate Client Reef Fish by Providing Tactile Stimulation." *Proceedings of the Royal Society of London, Series B: Biological Sciences* 268 (2001): 1495–501.

Bulliet, Richard W. *Hunters, Herders and Hamburgers: The Past and Future of Human-Animal Relationships*. New York: Columbia University Press, 2005.

Burbidge, Matthew. "The Big Beef." *Mail and Guardian*, 26 January to 1 February 2007.

Burdon Sanderson, John. "Inaugural Address to the Department of Physiology, BAAS." *Nature* 48 (1893): 464–72.

Burgdorf, Jeffrey, and Jaak Panksepp. "The Neurobiology of Positive Emotions." *Neuroscience and Biobehavioural Reviews* 30 (2007): 173–87.

———. "Tickling Induces Reward in Adolescent Rats. *Physiology & Behavior* 72 (2001): 167–73.

Burns, Katie. "North Carolina Shares a Model for Dealing with Disaster." *JAVMA News*, 1 April 2007. http://www.avma.org/onlnews/javma/apr07/070401h.asp (accessed 5 November 2009).

Byszewski, Elaine T. "Valuing Companion Animals in Wrongful Death Cases: A Survey of Current Court and Legislative Action and a Suggestion for Valuing Pecuniary Loss of Companionship." *Animal Law* 9 (2003): 215–41.

Cabanac Michel. "Physiological Role of Pleasure." *Science* 173 (1971): 1103–7.

Cabanac, Michel, and K. G. Johnson. "Analysis of a Conflict Between Palatability and Cold Exposure in Rats." *Physiology & Behavior* 31 (1983): 249–53.

Cadman, Mike. "Future Looks No Better for SA's 'Canned' Lions." *Sunday Independent*, 14 June 2009.

———. "New Law 'Won't Stop Canned-Lion Hunting in SA.'" *Sunday Independent*, 14 January 2007.

Carpenter, Edward. "The Need of a Rational and Humane Science." In *Humane Science Lectures*, 1– 34. London: George Bell & Sons, 1897.

———."The Science of the Future." In *Civilisation: Its Cause and Cure, and Other Essays*, 82–99. London: Swan Sonnenschein & Co., 1893.

———. *Vivisection: An Address to the Humanitarian League*. London: National Anti-Vivisection Society, 1903.

Carroll, Lewis. *The Annotated Alice: Alice's Adventures in Wonderland and Through the Looking Glass*. Illustrated by John Tenniel. 1865, 1871. With introduction and notes by Martin Gardner. Harmondsworth: Penguin Books, 1970.

———. "Vivisection as a Sign of the Times." *Pall Mall Gazette*, 12 February 1875.

Carruthers, Jane. *The Kruger National Park: A Social and Political History*. Pietermaritzburg: University of Natal Press, 1995.

"Cat Lovers Hold Memorial for 700 Culled Strays." *The Straits Times*, 9 June 2003.

Centers for Disease Control and Prevention, United States Department of Health and Human Services. "Animals in Public Evacuation Centers," http://www. bt.cdc.gov/disasters/animalspubevac.asp (accessed 5 November 2009).

Cheke, Anthony, S. "Establishing Extinction Dates—The Curious Case of the Dodo *Raphus cucullatus* and the Red Hen *Aphanapteryx bonasia*." *Ibis* 148 (2006): 155–58.

Cherrington, Lisa. *The People-Faces*. Wellington: Huia, 2004.

Chester, Jonathan. *The World of the Penguin*. San Francisco: Sierra Club, 1996.

"Chilling Out." *Tatler*, 5 April 2007.

Chris, *Watching Wildlife*, 154–5

City of New Orleans. "City Assisted Evacuation Plan (CAEP)." http://www. cityofno.com/Portals/NOHSEP/Resources/What%20is%20the%20CAEP.pdf (accessed 5 November 2009).

Clark, David. "On Being 'The Last Kantian in Nazi Germany': Dwelling with Animals after Levinas." In *Animal Acts: Configuring the Human in Western History*, edited by J. Ham and M. Senior, 165–97. New York: Routledge, 1997.

Clarke, Robert. *Ellen Swallow: The Woman Who Founded Ecology*. Chicago: Follett Publishing Company, 1973.

Cleland, John. *Experiment on Brute Animals*. London: J. W. Kolckmann, 1883.

Clout, M., and K. Ericksen. "Anatomy of a Disastrous Success: The Brushtail Possum as an Invasive Species." In *The Brushtail Possum: Biology, Impact and Management of an Introduced Marsupial*, edited by T. L. Montague, 1–9. Lincoln, NZ: Manaaki Whenua Press, 2000.

Clubb Ros, Marcus Rowcliffe, Phyllis Lee, Khyne U Mar, Cynthia Moss, and Georgia J Mason. "Compromised Survivorship in Zoo Elephants." *Science* 322 (2008): 1949.

Cobbe, Frances Power. "The Ethics of Zoophily," *Contemporary Review* 68 (1895): 497–508.

————. "The Scientific Spirit of the Age," *Contemporary Review* 54 (1888): 126–39.

Coe, Sue, and Judith Brody. *Sheep of Fools: A Song Cycle for Five Voices*. Seattle: Fantagraphics Books, 2005.

Coe, Sue. *Dead Meat*. New York: Four Walls Eight Windows, 1996.

————. *Pit's Letter*. New York: Four Walls Eight Windows, 2000.

Coetzee, J. M. "Meat Country," *Granta* 52 (1995): 42–52.

————. *The Lives of Animals*. Edited and introduced by Amy Gutmann. The University Center for Human Values Series. Princeton: Princeton University Press, 1999.

————. *The Lives of Animals*. London: Profile Books, 1999.

————. "Comments on Paola Cavalieri, 'A Dialogue on Perfectionism." In *The Death of the Animal: A Dialogue*, Paola Cavalieri with Matthew Calarco, J. M. Coetzee, Harlan B. Miller, and Cary Wolfe, 85–6. New York: Columbia University Press, 2009.

Coley, John D. "On the Importance of Comparative Research: The Case of Folkbiology." *Child Development* 71, no. 1 (2000): 82–90.

Collings, Matthew. *Art Crazy Nation: The Post-Blimey! Art World*. London: 21 Publishing, 2001.

Collingwood, R. G. *The Idea of History*. Oxford: Clarendon Press, 1946.

Collison, Lee-Shay and Natasha Prince. "Fugitive Baboon Evades Pursuers." *Cape Argus*, 4 April 2007.

Corr, Charles A. "Revisiting the Concept of Disenfranchised Grief." In Doka, *Disenfranchised Grief*, 39–60.

Crowell-Davis, Sharon L., Terry M. Curtis, and Rebecca J. Knowles. "Social Organization in the Cat: A Modern Understanding." *Journal of Feline Medicine and Surgery* 6 (2004): 19–28. http://zoopsy.free.fr/veille_biblio/social_organization_cat_2004.pdf (accessed 27 September 2009).

Csurhes, Steve. *An Assessment of the Potential Impact of Andropogon Gayanus (Gamba Grass) on the Economy, Environment and People of Queensland*. Brisbane: Queensland Department of Natural Resources, 2005.

"Cultural Conflict." *Cape Times*, 26 January 2007.

D'Antonio, C. M., and P. M. Vitousek. "Biological Invasions by Exotic Grasses, the Grass/Fire Cycle, and Global Change." *Annual Review of Ecology and Systematics* 23 (1992): 63–87.

Danbury T. C., C. A. Weeks, J. P. Chambers, A. E. Waterman-Pearson, and S. C. Kestin. "Self-Selection of the Analgesic Drug, Carprofen, by Lame Broiler Chickens." *Veterinary Record* 146 (2000): 307–11.

Darroch, Bob. *The Kiwi Who Lost His Mum*. Auckland: Reed Publishing, 2002.

Darwin, Charles. *The Expression of the Emotions in Man and Animals*. 1872. 3rd ed. Edited by Paul Ekman. London: HarperCollins, 1998.

Davis, Lucy. "Notes for a Singapore Bestiary: Gender, Sexuality, and Interspecies Exchanges in the City-State." In *documenta 12, Magazine no. 1–3 Reader*, edited by Georg Schöllhammer, Roger M. Buergel, and Ruth Noack, 366–79. Germany: Taschen Books, 2007.

————. "About FOCAS 6: Regional Animalities." 2007. http://www.substation. org/mag/features/about-focas-6-regional-animalities.html   (accessed   27 September 2009).

Davis, Matthew. "Saving New Orleans' Animals." *BBC News*, 10 September 2005. http://news.bbc.co.uk/2/hi/americas/4233408.stm   (accessed   5   November 2009).

Davis, Susan G. *Spectacular Nature: Corporate Culture and the Sea World Experience*. Berkeley: University of California Press, 1997.

Dawkins, Marian Stamp. "Behavioural Deprivation: A Central Problem in Animal Welfare." *Applied Animal Behaviour Science* 20 (1988): 209–25.

Day, David. *The Doomsday Book of Animals: A Unique History of Three Hundred Vanished Species*. London: Ebury Press, 1981.

Dayton, Leigh. "Climate: The Peril We Face." *The Weekend Australian*, 7–9 April 2007, main sec., p. 1.

Deeble, Mark, and Victoria Stone. "Kenya's Mzima Spring." *National Geographic* 200 (2001): 32–47.

Dehnhardt, Guido, Björn Mauck, Wolf Hanke, and Horst Bleckmann. "Hydrodynamic Trail-Following in Harbor Seals (*Phoca vitulina*)." *Science* 293 (2001): 29–31.

Deleuze, Gilles, and Félix Guattari. "Becoming-Intense Becoming-Animal." In *A Thousand Plateaus*, vol. 2 of *Capitalism and Schizophrenia*, 256–341. Minneapolis: University of Minnesota Press, 1987.

Den Hengst, Jan. *The Dodo: The Bird that Drew the Short Straw*. Marum, The Netherlands: Art Revisited, 2002.

Department of Agricultural Resources. Executive Office of Environmental Affairs. Commonwealth of Massachusetts. Emergency Order 2–AHO–05 Concerning Controls on Importation of Animals from Louisiana, Mississippi, and Alabama. 22 September 2005. http://www.mass.gov/agr/animalhealth/docs/Katrina-Emergency-Animal-Order-final.pdf (accessed 5 November 2009).

Department of Conservation and Land Management, State Government of Western Australia. *Encouraging Possums*. Wildlife Notes 6. Perth: Department of Conservation and Land Management, 1999.

Department of Conservation. Government of New Zealand. "Facts about Possums." http://www.doc.govt.nz/conservation/threats-and-impacts/animal-pests/animal-pests-a-z/possums/facts/ (accessed 28 October 2009).

————. "Possums," http://www.doc.govt.nz/conservation/threats-and-impacts/ animal-pests/animal-pests-a-z/possums/you-can-help/control-methods/ (accessed 17 July 2009).

Department of Natural Resources and Environment. State Government of Victoria. *Living with Possums*. Melbourne: State Government of Victoria, n.d.

Derrida, Jacques. "The Animal that Therefore I Am (More to follow)." Trans. David Wills, *Critical Inquiry* 28, Winter (2002): 396–418.

————. "'Eating Well,' or the Calculation of the Subject: An Interview with Jacques Derrida." In *Who Comes After the Subject?*, edited by Eduardo Cadava, Peter Connor, and Jean-Luc Nancy, 96–119. New York: Routledge, 1994.

Devlin, Erin. *As Kuku Slept*. Auckland: Raupo Publishing, 2006.

Diamond, Cora. "The Difficulty of Reality and the Difficulty of Philosophy." In *Philosophy and Animal Life,* edited by Stanley Cavell, Cora Diamond, John McDowell, Ian Hacking, and Cary Wolfe, 43–89. New York: Columbia University Press, 2008.

———. "Ethics, Imagination and the Method of Wittgenstein's *Tractatus*." In *Bilder der Philosophie*, edited by R. Heinrich, and H. Vetter, 55–90. Vienna: Oldenbourg, 1991.

Dissanayake, Ralph. "What did the Dodo Look Like?" *Biologist* 51, no. 3 (2004): 165–8.

Dodd, Lynley. *Slinky Malinki*. Wellington: Mallinson Rendel, 1990.

———. *Hairy Maclary from Donaldson's Diary*. Wellington: Mallinson Rendel, 1983.

Doka, Kenneth J., ed. Disenfranchised Grief: New Directions, Challenges, and Strategies for Practice. Champaign: Research Press, 2002.

Douglas, Mary. *Purity and Danger: An Analysis of Concepts of Pollution and Taboo*. London: Routledge and Kegan Paul, 1966.

Dowling, Tim. "My Life as a Penguin." *Guardian*, 15 November 2005, section G2.

Eason, C. T., B. Warburton, and R. Henderson. "Toxicants Used for Possum Control." In *The Brushtail Possum: Biology, Impact and Management of an Introduced Marsupial*, edited by T. L. Montague, 154–63. Lincoln, NZ: Manaaki Whenua Press, 2000.

Enigma of Kaspar Hauser, The. Directed by Werner Herzog. West Germany, Anchor Bay Entertainment, 1974.

"Environmental Thugs will Face Time in the Can." *Sunday Independent*, 7 May 2006.

Estep, Daniel W., David L. Lanier, and Donald A. Dewsbury. "Copulatory Behaviour and Nest Building Behaviour of Wild House Mice (*Mus musculus*)." *Animal Learning and Behavior* 3 (1975): 329–36.

Evans, Dylan. *Emotion: The Science of Sentiment*. Oxford: Oxford University Press, 2001.

"Facts Distorted in Yengeni Ritual Slaughter Drama." *Sunday Independent*, 28 January 2007.

Feh, Claudia, and Jeanne de Mazières. "Grooming at a Preferred Site Reduces Heart Rate in Horses." *Animal Behaviour* 46 (1993): 1191–4.

U.S. Department of Homeland Security, Federal Emergency Management Agency, Disaster Assistance Policy 9523.19 "Eligible Costs Related to Pet Evacuations and Sheltering," 24 October 2007.

———. "Information for Pet Owners." http://www.fema.gov/plan/prepare/animals.shtm (accessed 5 November 2009).

Fessler, Daniel M. T. "Shame in Two Cultures: Implications for Evolutionary Approaches." *Journal of Cognition and Culture* 4, no. 2 (2004): 207–62.

Fiddes, Nick. *Meat, A Natural Symbol*. London: Routledge, 1992.

Fisher, Philip. *Hard Facts: Setting and Form in the American Novel*. Oxford: Oxford University Press, 1985.

Fitzgerald, G. P., R. Wilkinson, and L. Saunders. "Public Perceptions and Issues in Possum Control." In *The Brushtail Possum: Biology, Impact and Management of an Introduced Marsupial*, edited by T. L. Montague, 187–96. Lincoln, NZ: Manaaki Whenua Press, 2000.

Flannery, Tim. *Beautiful Lies: Population and Environment in Australia*. Melbourne: Black Inc, 2003.

———. *A Gap in Nature: Discovering the World's Extinct Animals*. Melbourne: Text Publishing, 2001.

Forster, E. M. *A Passage to India*. Harmondsworth: Penguin, 2005.

Foster, Mary. "Superdome Evacuations Enter Second Day." 1 September 2005. All News Plus Wires (online database). Accession number 9/1/05 APDATASTREAM 12:30:23. Available from Westlaw. http://web2.westlaw.com.

Foster, Michael. "Vivisection." *Macmillan's Magazine* 29 (1874): 367–76.

Fox, Mem, and Julie Vivas. *Possum Magic*. 1983. 21st birthday edition. Malvern, South Australia: Omnibus Books, 2004.

Freeman, Carol. *Paper Tiger: A Visual History of the Thylacine*. Leiden: Brill, 2010.

French, Richard D. *Antivivisection and Medical Science in Victorian Society*. Princeton: Princeton University Press, 1975.

Fudge, Erica. *Animal*. London: Reaktion Books, 2002.

———. "The History of Animals." H-Animal: Ruminations. http://www.h-net.org/~animal/ruminations_fudge.html (accessed 4 April 2009).

———. "A Left-Handed Blow: Writing the History of Animals." In *Representing Animals*, edited by Nigel Rothfels, 3–18. Bloomington: University of Indiana Press, 2002.

Fuller, Errol. *Dodo: From Extinction to Icon*. London: Collins, 2002.

Galdikas, Birute. "Living with Orangutans." *National Geographic* 157, no. 6 (1980): 830–53.

Garber, Marjorie. *Academic Instincts*. Princeton: Princeton University Press, 2001.

Gates, Barbara T. *Kindred Nature: Victorian and Edwardian Women Embrace the Living World*. Chicago: University of Chicago Press, 1998.

Geison, Gerald. *Michael Foster and the Cambridge School of Physiology*. Princeton: Princeton University Press, 1978.

Gillmor, Alison. "Happy Feet Awkward, Flightless Bird of a Film." *Winnipeg Free Press*, 17 November 2006.

Glass, Alexi. *Lisa Roet*. Fishermans Bend, Melbourne: Craftsman House, 2004.

Goodall, Jane. *In the Shadow of Man*. Boston: Houghton Mifflin, 1971.

Gooding, David. "Of Dodos and Dutchmen: Reflections on the Nature of History." *Critical Quarterly* 47, no. 4 (2005): 32–47.

———. "Putting the Agency Back into Experiment." In *Science as Practice and Culture*, edited by A. Pickering, 65–111. Chicago: University of Chicago Press, 1992.

Gophe, Myolisi. "Abbatoir Workers Have No Beef with Tradition." *Weekend Argus*, 27 January 2007.

———. "Yengeni's Spear Starts Two-Day Welcome Party." *Weekend Argus*, 20 January 2007.

Gordon, Cosmo Duff. "Primate Scream." *Sunday Times*, 9 September 2007, Lifestyle sec.

Gordon, Neta. "Sign and Symbol in Barbara Gowdy's *The White Bone.*" *Canadian Literature* 185 (2005): 76–91.

Gosling, Melanie. "Banned Lethal Poison Led to Deaths of Baboons." *Cape Times*, 23 August 2006.

———. "Poisoned 'Baboon Woman' Jenni Trethowan Suffers from Relapses." *Cape Times*, 14 November 2006.

Gould, Peter C. *Early Green Politics: Back to Nature, Back to the Land, and Socialism in Britain, 1880–1900*. New York: St. Martin's Press, 1988.

Gould, Stephen J. "The Dodo in the Caucus Race," *Natural History* 105, no. 11 (1996): 22–33.

Gowdy, Barbara, *The White Bone*. New York: Picador, 1998.

Graham, Julia. *Opo the Happy Dolphin*. Auckland: Golden Press, 1979.

Green, W. *The Use of 1080 for Pest Control: A Discussion Document*. Wellington: Animal Health Board and Department of Conservation, 2004.

Griffin, Donald R. *Animal Minds*. Chicago: University of Chicago Press, 1992.

———. *Animal Thinking*. Harvard: Harvard University Press, 1984.

———. *The Question of Animal Awareness*. New York: Rockefeller University Press, 1976.

Grizzly Man. Directed by Werner Herzog. Santa Monica, CA: Lion Gate Films/ Discovery Docs, 2005.

Grove, Richard H. *Green Imperialism: Colonial Expansion, Tropical Island Edens and the Origins of Environmentalism, 1600–1800*. Cambridge: Cambridge University Press, 1995.

Guattari, Félix. *Chaosmosis: An Ethico-Aesthetic Paradigm*. Translated by Paul Bains and Julian Pefanis. Sydney: Power Publications, 1992.

Halberstam, Judith. "Animating Revolt/Revolting Animation: Penguin Love, Doll Sex and the Spectacle of the Queer Nonhuman." In *Queering the Non/Human*, edited by Noreen Giffney, and Myra J. Hird, 265–81. Aldershot: Ashgate, 2008.

Happy Feet. Directed by George Miller. Burbank, CA: Warner Bros. Pictures, 2006.

Haraway, Donna J. *When Species Meet*. Minneapolis: University of Minnesota Press, 2008.

Harding, Luke. "Females Flown in to P-p-p-pick Up 'Gay' Penguins." *Guardian*, 15 February 2005. http://www.guardian.co.uk/germany/article/0,2763,1414804,00.html (accessed 25 May 2009).

Hatch, Anita, G. S. Wiberg, Tibor Balazs, and H. C. Grice. "Longterm Isolation Stress in Rats." *Science* 142 (1963): 507.

Hearne, Vicki. *Adam's Task: Calling Animals By Name*. New York: Skyhorse Publishing, 2007.

Heng, Geraldine, and Devan, Janadas. "State Fatherhood: The Politics of Nationalism, Sexuality and Race in Singapore." In *Nationalisms and Sexualities*, edited by Andrew Parker, Mary Russo, Doris Sommer, and Patricia Yaeger, 343–64. New York: Routledge, 1992.

Herzog, *The Enigma of Kaspar Hauser*

Hessler, Kathy. "Mediating Animal Law Matters." *Journal of Animal Law and Ethics* 2 (2007): 21–75.

Heyers, Dominik, Martina Manns, Harald Luksch, Onur Güntürkün, and Henrik Mouritsen. "A Visual Pathway Links Brain Structures Active During Magnetic Compass Orientation in Migratory Birds." *PLoS ONE* 2 (2007): e937.

Hill, N. J., K. A. Carbery, and E. M. Deane. "Human-Possum Conflict in Urban Sydney, Australia: Public Perceptions and Implications for Species Management." *Human Dimensions of Wildlife* 12 (2007): 101–13.

Hofstatter, Amrei. "My Dearest, Dearest Creature." *Belio*, no. 27 (2008): 25–36. http://www.angelasinger.com/?page_id=7 (accessed 28 September 2009).

Holden, Stephen. "A Reprieve for Reality in New Crop of Films." *The New York Times*. September 2, 2005, http://www.nytimes.com/2005/09/02/movies/02note/html (accessed May 25, 2009).

Horwitz, Jane. "The Family Filmgoer: Watching with Kids in Mind." Review of *Happy Feet* (Warner Bros movie). *Washington Post*, 17 November 2006, http://www.washingtonpost.com/wp-dyn/content/article/2006/11/16/AR2006111600269.html (accessed June 1, 2009).

Huisman, Bienne, Nashira Davids, Ndivhulo Mafela, and Buyekezwa Makwabe. "Yengeni Hosts Cleansing Party." *Sunday Times*, 21 January 2007.

Human Rights Watch. "World Report Chapter: Singapore." 2009. http://www.hrw.org/en/node/79247 (accessed 27 September 2009).

Hume, Julian P. "The History of the Dodo *Raphus cucullatus* and the Penguin of Mauritius." *Historical Biology* 18, no. 2 (2006): 65–89.

Hutton, Ian. *Lord Howe Island.* Canberra: Conservation Press, 1986.

Jamieson, "Does Philosophy Have Anything New to Say About Animals?" (Discussion panel, Minding Animals: International Conference, Newcastle, NSW, Australia, 13–18 July 2009).

Janis, C. M., J. Damuth, and J. M. Theodor. "Miocene Ungulates and Terrestrial Primary Productivity: Where Have All the Browsers Gone?" *Proceedings of the National Academy of Sciences of the United States of America* 97, no. 14 (2000): 7899–904.

Jaschinski, Britta. *Wild Things.* London: Thames and Hudson, 2003.

———. *Zoo.* London: Phaidon, 1996.

Johnson, Chris. *Australia's Mammal Extinctions: A 50,000 Year History.* Melbourne: Cambridge University Press, 2006.

Jones, Rachel "Value of Not Knowing." symposium *On Not Knowing: How Artists Think.*

Jones, Rhys. "Fire-Stick Farming." *Australian Natural History* 16 (1969): 224–28.

Kallir, Jane. "Sue Coe: *Sheep of Fools*." New York: Galerie St. Etienne, 2005. An exhibition leaflet.

Kalof, Linda, and Amy Fitzgerald, eds. *The Animals Reader: The Essential Classic and Contemporary Writings.* Oxford and New York: Berg, 2007.

"Katrina's Animal Rescue." *Nature.* Thirteen/WNET New York and National Geographic Television, 20 November 2005.

Kean, Hilda. *Animal Rights: Political and Social Change in Britain Since 1800.* London: Reaktion Books, 1998.

Kelly, Christopher. "Waddle and Twaddle." *Weekend Australian*, September 3–4, 2005.

Kemp, Yunus. "Murder Most Fowl: Theatre Fans Clucking Mad after Baxter Bloodbath." *Cape Argus*, 11 August 2003.

Kenyon-Jones, Christine. *Kindred Brutes: Animals in Romantic Period Writing.* Aldershot: Ashgate, 2001.

Keverne, Eric B. "Primate Social Relationships: Their Determinants and Consequences." *Advances in the Study of Behaviour* 21 (1992): 1–36.

"Khayelitsha Commemorates World Farm Animal Day—African Style." *Animal Voice* (Summer 2001/2): 3.

King, Margaret J. "The Audience in the Wilderness: The Disney Nature Films." *Journal of Popular Film and Television* 24, no. 2 (Summer 1996): 60–68.

King, Thomas. *Green Grass, Running Water.* New York: Houghton Mifflin, 1993.

Kingsford, Anna, "Unscientific Science: Moral Aspects of Vivisection." In *Spiritual Therapeutics,* edited by W. Colville, Jr., 292–308. Edinburgh: 1883.

Kitchener, Andrew. "Justice at Last for the Dodo." *New Scientist*, 28 August 1993, 24–7.

———. "On the External Appearance of the Dodo, *Raphus Cucullatus* (L., 1758)." *Archives of Natural History* 20, no. 2 (1993): 270–301.

Kiwi Conservation Club. "Possums." http://www.kcc.org.nz/pests/possum.asp (accessed 17 July 2009).

Klotz, Hattie. "Fur Fashion to the Rescue." First published in *The Ottawa Citizen*, March 8, 2001. http://www.maninnature.com/AFibres/Trapping/Trapping1b.html (accessed 17 July 2009).

Knabb, Richard D., Jamie R. Rhome, and Daniel P. Brown. "Tropical Cyclone Report: Hurricane Katrina, August 23–30, 2005," 20 December 2005. http://www.nhc.noaa.gov/pdf/TCR-AL122005_Katrina.pdf (accessed 5 November 2009).

Knight, John. "Maternal Feelings on Monkey Mountain." In *Mixed Emotions: Anthropological Studies of Feeling*, edited by Kay Milton and Maruška Svašek, 179–93. Oxford: Berg, 2005.

———. "On the Extinction of the Japanese Wolf." *Asian Folklore Studies* 56 (1997): 129–59.

Kua, Dawn. "David Hooi Sentenced." http://catwelfare.blogspot.com/2006/09/david-hooi-sentenced.html (accessed 27 September 2009).

La Marche de l'Empereur. Directed by Luc Jacquet. Paris: Bonne Pioche, 2005.

"Lady Lavona's Cabinet of Curiosities," http://ladylavona.blogspot.com/.

Langford, Dale J, Sara E. Crager, Zarrar, Shehzad, Shad B. Smith, Susana G. Sotocinal, Jeremy S. Levenstadt, Mona Lisa Chanda, Daniel J. Levitin, and Jeffrey S. Mogil. "Social Modulation of Pain as Evidence for Empathy in Mice." *Science* 312 (2006): 1967–70.

Lankester, Edwin Ray. "Vivisection." *Nature* 9 (1873): 145.

Lansbury, Coral. *The Old Brown Dog: Women, Workers and Vivisection in Edwardian England.* Madison: University of Wisconsin Press, 1985.

Laplanche, Jean, and Jean-Bertrand Pontalis. *The Language of Psychoanalysis.* Translated by Donald Nicholson-Smith. London: Karnac Books, 1973.

Latour, Bruno. *Politics of Nature: How to Bring the Sciences into Democracy.* Translated by Catherine Porter. Cambridge, MA: Harvard University Press, 2004.

Lauraallen, "Are Government Officials Ready to Evacuate and Shelter Animals in Disasters?" Animal Law Coalition, online posting, 29 August 2008, http://www.animallawcoalition.com/animals-and-politics/article/580.

Lawrence, Richard D. "Primer on Wrongful Death Claims." *Trial* 40, no. 2 (2004): 42–7.

Lee-Johnson, Eric, and Elizabeth Lee-Johnson. *Opo the Hokianga Dolphin.* Auckland: David Ling, 1994.

Lee, Vernon. "Vivisection: An Evolutionist to Evolutionists." *The Contemporary Review* 41 (1882): 788–811.

Levinas, Emmanuel. "The Name of a Dog, or Natural Rights." In *Difficult Freedom*, translated by S. Hand, 151–3. London: The Athlone Press, 1990.

———. "The Paradox of Morality: An Interview with Emmanuel Levinas." In *The Provocation of Levinas: Rethinking the Other*, edited by R. Bernasconi and D. Wood, translated by A. Benjamin and T. Wright, 168–80. London: Routledge, 1988.

———. *Totality and Infinity.* Translated by Alphonso Lingis. Pittsburgh: Duquesne University Press, 1969.

Lewis, Jeremy. *Penguin Special: The Life and Times of Allen Lane.* London: Viking, 2005.

Lippincott, Louise, and Andreas Blühm. *Fierce Friends: Artists and Animals 1750–1900.* London: Merrell, 2005.

Loeffler, K. "Pain and Suffering in Animals." *Berliner und Münchener Tierärztliche Wochenschrift* 103 (1990): 257–61.

Long, Tom. "'Happy Feet' Will Leave You Spinning." *The Detroit News*, 17 November 2006.

Low, Tim. "Born to Burn." *Nature Australia* 28, no. 1 (2004): 24–25.

———. *Climate Change and Invasive Species: A Review of Interactions.* Canberra: Biological Diversity Advisory Committee, 2008.

———. *Feral Future: The Untold Story of Australia's Exotic Invaders.* Melbourne: Penguin, 1999.

———. *The New Nature: Winners and Losers in Wild Australia.* Melbourne: Penguin, 2002.

"Luck for Plucky Clucker." *Cape Argus*, 11 August 2003.

Lulka, David. "Consuming Timothy Treadwell: Redefining Nonhuman Agency in Light of Herzog's Grizzly Man." In McFarland and Hediger, *Animals and Agency*, 67–87.

Macleod, Fiona. "The Feud over Frida." *Mail and Guardian*, 19–25 January 2007.

Magel, Charles R. *Keyguide to Information Sources in Animal Rights,* Jefferson, NC: McFarland, 1989.

Malamud, Randy. *Reading Zoos: Representations of Animals and Captivity*. New York: New York University Press, 1998.

March of the Penguins. Directed by Luc Jacquet. American edition. Los Angeles: National Geographic Films, 2005.

Marion, Nancy E. "Symbolic Policies in Clinton's Crime Control Agenda." *Buffalo Criminal Law Review* 1 (1997): 67–108.

Martel, Yann. *Life of Pi*. Toronto: Random House, 2001.

Martin, Garry. "The Phrase Finder: Dead as a Dodo." http://www.phrases.org.uk/meanings/38900.html (accessed 31 January 2008).

Martin, Stephen. *Penguin*. London: Reaktion, 2009.

Marvin, Garry. "Wild Killings: Contesting the Animal in Hunting." In Animal Studies Group, *Killing Animals*, 10–29.

Masson, Jeffrey Moussaieff. *The Pig Who Sang to the Moon: The Emotional World of Farm Animals*. New York: Ballantine, 2003.

Matthews, A., D. Lunney, K. Waples, and J. Hardy. "Brushtail Possums: 'Champion of the Suburbs' or 'Our Tormentors'?" In *Urban Wildlife: More than Meets the Eye*, edited by D. Lunney and S. Burgin, 159–68. Mosman, NSW: Royal Zoological Society of New South Wales, 2004.

Mayer, Jed. "Ruskin, Vivisection, and Scientific Knowledge." *Nineteenth-Century Prose* 35, no. 1 (Spring 2008): 200–222.

Mayo, Jenny. "Lovable 'Happy Feet' Goes Adrift." *Washington Times*, 17 November 2006.

McFarland, Sarah E. "Dancing Penguins and a Pretentious Raccoon: Animated Animals and 21st Century Environmentalism." In *Animals and Agency*, 89–103.

———, and Ryan Hediger, eds. *Animals and Agency*. Leiden: Brill, 2009.

McGlashan, Don. "Miracle Sun." *Warm Hand*. Auckland: Arch Hill Records, 2006.

McKibben, Bill. *The End of Nature*. London: Penguin, 1990.

McNeill, Bob. "Opo Fifty Years On." *TV3 News*. Auckland: TV3, February 2006.

Miller, Jonathan. "March of the Conservatives: Penguin Film as Political Fodder." *New York Times*, 13 September 2005. http://www.nytimes.com/2005/09/13/science/13peng.html?ei=5088&en=36effea48de3fa22&ex=1284264000&partner=rssnyt&emc=rss&pagewanted=print (accessed 30 May 2009).

Miller, K. K., P. R. Brown and I. Temby. "Attitudes Towards Possums: A Need for Education?" *The Victorian Naturalist* 116, no. 4 (1999): 120–26.

Milton, Kay. "Anthropomorphism or Egomorphism?: The Perception of Non-Human Persons by Human Ones." In *Animals in Person: Cultural Perspectives on Animal-human Intimacy*, edited by John Knight, 255–71. Oxford: Berg, 2005.

Mitchell, W. J. T. "The Rights of Things." In Wolfe, *Animal Rites*, ix–xiv.

Mitman, Gregg. *Reel Nature: America's Romance with Wildlife on Film.* Cambridge, MA: Harvard University Press, 1999.

Morris, William. *Collected Works.* Edited by May Morris. 24 vols. London: Longmans, Green and Co., 1910–15.

Morris, Desmond. "Foreword." In *Fierce Friends: Artists and Animals 1750–1900*, Louise Lippincott and Andreas Blüm, 9–11. London: Merrell, 2005.

Morse, Deborah Denenholz, and Martin A. Danahy, eds. *Victorian Animal Dreams: Representations of Animals in Victorian Literature and Culture.* Aldershot, Hampshire: Ashgate, 2007.

Movietone News #47. Directed by Rudall Hayward. Wellington: National Film Unit, 1956.

Moya, Fikile-Ntsikelelo. "SPCA Needs to Work with Black People." *Mail and Guardian*, 26 January –1 February 2007.

Mulhall, Stephen. *The Wounded Animal: J. M. Coetzee and the Difficulty of Reality in Literature and Philosophy.* Princeton: Princeton University Press, 2009.

Muller, Max. "The Science of Language." In *Nineteenth Century Science: A Selection of Original Texts*, edited by A. S. Weber, 266–77. Peterborough, Ontario: Broadview Press, 2000.

Murphy, Patrick D. "'The Whole Wide World Was Scrubbed Clean': The Androcentric Animation of Denatured Disney." In *From Mouse to Mermaid: The Politics of Film, Gender, and Culture*, edited by Elizabeth Bell, Lynda Haas, and Laura Sells, 125–36. Bloomington: Indiana University Press, 1995.

Mutch, Robert W. "Wildland Fires and Ecosystems—A Hypothesis" *Ecology* 51, no. 6 (1970): 1046–51.

Nattrass, Ric. *Talking Wildlife.* Archerfield, QLD: Steve Parish Publishing, 2004.

North Carolina State Animal Response Team.. "SART: State Animal Response Teams." http://nc.sartusa.org (accessed 5 November 2009).

"Nuisance Stray Cats Will Be Caught, Put Down." *The Straits Times*, 24 May 2003.

Nussbaum, Martha C. *Frontiers of Justice: Disability, Nationality, Species Membership.* Cambridge: Belknap Press of Harvard University Press, 2006.

O'Dowd, Dennis J., Peter T. Green, and P. S. Lake. "Invasional Meltdown on an Oceanic Island." *Ecology Letters* 6 (2003): 812–17.

Opononi: The Town that Lost a Miracle. Directed by Barry Barclay and James McNeish. Wellington: Pacific Films, 1972.

Osborne, Colin P. "Atmosphere, Ecology and Evolution: What Drove the Miocene Expansion of C-4 Grasslands?" *Journal of Ecology* 96, no. 1 (2008): 35–45.

Ouida [Marie Louise Ramé]. "The Ugliness of Modern Life." Chap. 10 in *Critical Studies*, 212–36. London: T. Fisher Unwin, 1900.

Owen, Alwyn and Dave Gunson. *How the Kiwi Lost its Wings*. Auckland: Reed Publishing, 2002.

Owen, Richard. "On the Dodo. Part II: Notes on the Articulated Skeleton of the Dodo (*Didus ineptus*, Linn.) in the British Museum." *Transactions of the Zoological Society of London* 7, no. 3 (1872): 513–25.

Panksepp, Jaak. *Affective Neuroscience*. Oxford: Oxford University Press, 1998.

Panksepp, Jaak, and Jeffrey Burgdorf. "'Laughing' Rats and the Evolutionary Antecedents of Human Joy?" *Physiology & Behavior* 79 (2003): 533–47.

Panksepp, Jules B., and Robert Huber. "Ethological Analyses of Crayfish Behavior: A New Invertebrate System for Measuring the Rewarding Properties of Psychostimulants." *Behavioural Brain Research* 153 (2003): 171–80.

Pechey, Laura Charlotte. "Problem Animals, Indigeneity and Land: The Chacma Baboon in South African Writing." *Current Writing: Text and Reception in Southern Africa* 18, no. 1 (2006): 42–60.

Perkins, David. *Romanticism and Animal Rights*. New York: Cambridge University Press, 2003.

Persinger, M. A. "Rats' Preferences for an Analgesic Compared to Water: An Alternative to "Killing the Rat So It Does Not Suffer." *Perceptual and Motor Skills* 96 (2003): 674–80.

Peters, Melanie. "SPCA Finds No Evidence of Cruelty During Yengeni Ritual." *Weekend Argus* 27 January 2007.

Philanthropos [pseud.]. *Physiological Cruelty; or, Fact vs. Fancy. An Inquiry into the Vivisection Question.* London: Tinsley Bros., 1883.

Philo, Chris, and Chris Wilbert, *Animals Spaces, Beastly Places: New Geographies of Human-Animal Relations*. London: Routledge, 2000.

Pickover, Michele. *Animal Rights in South Africa.* Cape Town: Double Storey, 2005.

Pickstone, John V. "Science in Nineteenth-Century England: Plural Configurations and Singular Politics" In *The Organisation of Knowledge in Victorian Britain*, edited by Martin Daunton, 29–60. Oxford: Oxford University Press, 2005.

———. *Ways of Knowing: A New History of Science, Technology, and Medicine*. Chicago: University of Chicago Press, 2001.

Plumwood, Val. "The Concept of a Cultural Landscape." *Ethics and the Environment* 11, no. 2 (2006): 115–50.

———. "Prey to a Crocodile." *Aisling Magazine* 30 (2002): 1–5. http://www. aislingmagazine.com/aislingmagazine/artices/TAM30/valPlumwood.html (accessed 26 September 2009).

Pollan, Michael. *The Botany of Desire: A Plant's Eye View of the World*. New York: Random House, 2001.

Pollock, Mary Sanders, and Catherine Rainwater, eds. *Figuring Animals: Essays on Animal Images in Art, Literature, Philosophy and Popular Culture*. New York: Palgrave Macmillan, 2005.

Potts, Geoffrey W. "The Ethology of *Labroides dimidiatus* on Aldabra." *Animal Behaviour* 21 (1973): 250–91.

Primatt, Humphrey. *A Dissertation on the Duty of Mercy and the Sin of Cruelty to Brute Animals*. 1776. American Libraries digital archive. http://www.archive.org/details/adissertationon00primgoog.

"Production Notes: *Happy Feet.*" *Visual Hollywood*. http://www.visualhollywood.com/movies/happy-feet/notes.pdf (accessed 25 May 2009).

Protectors of Public Lands Victoria. http://www.protectorsofpubliclandsvic.com/ (accessed 1 May 2007; page now discontinued).

Pyne, Stephen J. *Burning Bush: A Fire History of Australia*. New York, NY: Henry Holt and Company, 1991.

Quammen, David. *The Song of the Dodo: Island Biogeography in the Age of Extinctions*. London: Hutchinson, 1996.

Rae, Paul and Low, Kee Hong. "Nosing Around: A Singapore Scent Trail." *Performance Research* 8, no. 3 (2003): 44–54.

Rae, Paul. "Cat's Entertainment: Feline Performance in the Lion City." *TDR/The Drama Review* 51, no. 1 (2007): 119–37.

Ranger, Terence. *Voices From the Rocks: Nature, Culture and History in the Matapos Hills of Zimbabwe*. Bloomington: Indiana University Press, 1999.

Ranlett, John. "'Checking Nature's Desecration': Late-Victorian Environmental Organization." *Victorian Studies* 26 (1983): 198–222.

Regan, Tom, and Peter Singer. *Animals Rights and Human Obligations*. Englewood Cliffs: Prentice Hall, 1976.

Regan, Tom. *The Case for Animal Rights*. Berkeley, CA: University of California Press, 1983.

Reporters Without Borders. "Press Freedom Index 2009." 2009. http://www.rsf.org/IMG/pdf/classement_en.pdf (accessed 26 October 2009).

"Researchers Find Gay Penguins in Japanese Aquariums." *Agence France Presse*, December 25, 2004.

Ritvo, Harriet. *The Animal Estate: The English and Other Creatures in the Victorian Age*. Cambridge: Harvard University Press, 1987.

———. "History and Animal Studies." *Society and Animals* 10, no. 4 (2002): 403–6.

———. "Manchester v. Thirlmere and the Construction of the Victorian Environment." *Victorian Studies* 49, no. 3 (2007): 457–81.

———. *The Platypus and the Mermaid, and Other Figments of the Classifying Imagination*. Cambridge, MA: Harvard University Press, 1997.

Rollin, Bernard. "Scientific Ideology, Anthropomorphism, Anecdote and Ethics." *New Ideas in Psychology* 13 (2000): 109–18.

Root, William C. "'Man's Best Friend': Property or Family Member? An Examination of the Legal Classification of Companion Animals and Its Impact on Damages Recoverable for Their Wrongful Death or Injury." *Villanova Law Review* 47 (2002): 423–49.

Rossiter, N. A., M. M. Douglas, S. A. Setterfield, and L. B. Hutley. "Testing the Grass-Fire Cycle: Alien Grass Invasion in the Tropical Savannas of Northern Australia." *Diversity and Distributions* 9 (2003): 169–76.

Rothfels, Nigel, ed. *Representing Animals*. Bloomington: Indiana University Press, 2002.

Ruskin, John. *The Works of John Ruskin*. 39 vols. Edited by E. T. Cook and Alexander Wedderburn. London: G. Allen, 1903–12.

Rutherford, William. "Opening Address to the Department of Anatomy and Physiology, BAAS." *Nature* 9 (1873): 455–57.

Salt, Henry S. "Vivisection (Reprinting of October, 1903 Letter to *Daily News*)." *Humanity* 20 (1903): 155–6.

Sapa. "Yengeni Row Political Bull, says SPCA." *Cape Argus*, 14 February 2007.

"Hurricane Preparedness: One-Third on High Risk Coast Will Refuse Evacuation Order, According to Survey." *Science Daily*, 26 July 2007. http://www.sciencedaily.com/releases/2007/07/070724113927.htm (accessed 5 November 2009).

Scott, A. C. "The Pre-Quaternary History of Fire." *Palaeogeography, Palaeoclimatology, Palaeoecology* 164, no. 1–4 (1997): 281–329.

Scott, Jane. "Animal— Unsettled Boundaries." In *Unsettled Boundaries*, 4–5.

Selye, Hans. *Stress Without Distress*. Philadelphia: J B Lippincott Co, 1974.

Senate Committee on Homeland Security and Governmental Affairs, *Hurricane Katrina: A Nation Still Unprepared*, 109th Cong., 2d sess., 2006, S. Rep. 109–322, 261.

Serpell, James A. "Anthropomorphism and Anthropomorphic Selection—Beyond the 'Cute Response.'" *Society and Animals* 11, no. 1 (2003): 83–100.

Shadbolt, Maurice. "Opo." Chap. 15 in *Love and Legend: Some Twentieth Century New Zealanders*. Auckland: Hodder and Stoughton, 1976.

———. *This Summer's Dolphin*. London: Cassell, 1969.

Shakespeare, William. *The Tempest*. Edited by Stephen Orgel. Oxford: Oxford University Press, 1998.

Shapiro, Kenneth Joel. *Human-Animal Studies: Growing the Field, Applying the Field*. Animals and Society Institute Policy Paper. Ann Arbor: Animals and Society Institute, 2008.

Sharp Jody L., Timothy G. Zammit, Toni A. Azar and David M. Lawson. "Stress-Like Responses to Common Procedures in Male Rats Housed Alone or with Other Rats." *Contemporary Topics in Laboratory Animal Science* 41 (2002): 8–14.

Sherman, Arloc, and Isaac Shapiro. "Essential Facts about the Victims of Hurricane Katrina." Center on Budget and Policy Priorities, 19 September 2005. http://www.cbpp.org/9-19-05pov.pdf (accessed 5 November 2009).

Sherwin, Chris M., and I. A. S. Olsson. "Housing Conditions Affect Self-Administration of Anxiolytic by Laboratory Mice." *Animal Welfare* 13 (2004): 33–9.

Shirihai, Hadoram. *A Complete Guide to Antarctic Wildlife: The Birds and Marine Mammals of the Antarctic Continent and Southern Ocean*. Degerby: Alula Press, 2002.

Silverman, Jerald. "Sentience and Sensation." *Lab Animal* 37 (2008): 465–67.

Simmons, Rebecca. "No Pet Left Behind: The PETS Act Calls for Disaster Plans to Include Animals." Humane Society of the United States, 20 April 2006. http://www.hsus.org/pets/pets_related_news_and_events/no_pet_left_behind_the_pets.html (accessed 5 November 2009).

Sinclair, Upton. *The Jungle*. London: T. Werner Laurie Ltd, 1906.

Singer, Peter. *Animal Liberation*. 2nd ed. New York: Avon, 1990.

————. "Preface." In *Animal Philosophy*, edited by Mathhew Calarco and Peter Atterton, xi–xiii. London: Continuum, 2004.

Skatssoon, Judy. "Civet Cats Not the Original Source of SARS." *ABC Science Online*, 5 October 2004. http://www.abc.net.au/science/news/stories/2004/1213117.htm (accessed 27 September 2009).

Slavick, Elin. "Art◇Activism: An Interview with Sue Coe." *MediaReader Quarterly* 4 (2000). http://www.mediareader.org/Issue4Stories/4_SueCoe.html (accessed 28 September 2009).

Smee, Sebastian. "Beastly Goings On." *The Weekend Australian*, 10–11 March 2007, Review sec.

Smith, David. "Film about Penguins' Tough Life Inspires America's Religious Right." *Guardian Weekly*, 23–29 September 2005.

Spivak, Gayatri Chakravorty. "Can the Subaltern Speak?" In *Marxism and the Interpretation of Culture*, edited by Cary Nelson and Larry Grossberg, 271–313. Urbana: University of Illinois Press, 1988.

Stauffer, Robert C. "Haeckel, Darwin, and Ecology." *The Quarterly Review of Biology* 32 (1957): 138–45.

Stephen, Leslie. "Thoughts of an Outsider: The Ethics of Vivisection." *The Cornhill Magazine* 33 (1876): 468–78.

Stonehouse, Bernard. *Penguins*. London: Barker, 1968.

Strauss, Valerie. "Humanized 'Penguins'? That Idea Just Doesn't Fly." *Washington Post*, 27 November 2005.

Strickland, H. E., and A. G. Melville. *The Dodo and its Kindred; or the History, Affinities, and Osteology of the Dodo, Solitaire, and other Extinct Birds of the Islands of Mauritius, Rodriguez, and Bourban*. London: Reeve, Benham and Reeve, 1848.

Stuff.co.nz. http://www.stuff.co.nz/stuff/0,2106,2954125a11,00.html (accessed 20 June 2004; page discontinued).

Tame, Adrian. "Gay Old Time Over a Little Fairy Bird." *Sunday Herald Sun*, 16 April 2006.

Tannenbaum, Jerrold. "The Paradigm Shift Toward Animal Happiness: What it is, Why it is Happening, and What It Portends for Medical Research." In *Why Animal Experimentation Matters: The Use of Animals in Medical Research*, edited by Frankel Paul Ellen and Paul Jeffrey, 93–130. Transaction Publishers: New York, 2001.

Tebbich, Sabine, Redouan Bshary, and Alexandra S. Grutter. "Cleaner Fish *Labroides dimidiatus* Recognise Familiar Clients." *Animal Cognition* 5 (2002): 139–45.

Temby, Ian. *Wild Neighbours: The Humane Approach to Living with Wildlife.* Broadway, NSW: Citrus Press, 2005.

Thomas, Keith. *Man and the Natural World: Changing Attitudes in England 1500–1800.* London: Allen Lane, 1983.

Todorov, Tzvetan. *The Conquest of America: The Question of the Other.* New York: Harper and Row, 1984.

Toi, Piwai. "Opo the Gay Dolphin." *Te Ao Hou* 23 (1958): 22–4.

"Transcript of National Day Rally 2009 Speech (Part 3 – Religious Harmony)." 2009. http://www.news.gov.sg/public/sgpc/en/media_releases/agencies/pmo/transcript/T-20090816-2.html (accessed 27 September 2009).

Turney, Samuel, and Anthony S. Cheke, "Dead as a Dodo: The Fortuitous Rise to Fame of an Extinction Icon." *Historical Biology* 20, no. 2 (2008): 149–63.

Tushnet, Mark, and Larry Yackle. "Symbolic Statutes and Real Laws: The Pathologies of the Antiterrorism and Effective Death Penalty Act and the Prison Litigation Reform Act." *Duke Law Journal* 47 (1997): 74–86.

Tvveede Boeck, Het. Journael oft Dagh-register/inhoudende een warachtig verhael ende historische vertellinghe vande reyse/gedaen door de acht schepen van Amstelredamme... Amsterdam: Cornelius Claesz, 1601.

Unsettled Boundaries. Edited by Jane Scott and Linda Williams. Melbourne: Melbourne International Arts Festival, 2006. An exhibition program.

Van Loo, Pascale L. P., Arco C. de Groot, Bert F. M. Van Zutphen, and Vera Bauman. "Do Male Mice Prefer or Avoid Each Other's Company? Influence of Hierarchy, Kinship, and Familiarity." *Journal of Applied Animal Welfare Science* 4 (2001): 91–103.

Van Riel, Fransje. "Cub at Mercy of Canned Hunting Industry." *Cape Times*, 2 March 2007.

———. *Life with Darwin and Other Baboons.* London: Penguin, 2003.

Varela, Francisco J. *Ethical Know-How: Action, Wisdom, and Cognition.* Stanford, CA: Stanford University Press, 1999.

Vaughn, Susan. "Staying True to a Unique Vision of Art." *Los Angeles Times*, 1 April 2001. http://graphicwitness.org/coe/latimes.htm (accessed 28 September 2009).

Verbaan, Aly. "David the Baboon Darted in Claremont." *People's Post*, 10 April 2007.

Voiceless: I Feel Therefore I Am. Edited by Charles Green and Ondine Sherman. Sydney: Sherman Galleries, 2007. An exhibition catalogue.

Walker, Ruth. "Scholars Probe Changing Legal, Cultural Status of Animals." *Havard Gazette Online*, 17 May 2007. http://www.news.harvard.edu/gazette/2007/05.17/10-crossings.html.

Wasko, Janet. *Understanding Disney: The Manufacture of Fantasy.* Cambridge: Polity Press, 2001.

Weyler, Rex. *Song of the Whale.* New York: Anchor Press/Doubleday, 1986.

Whatmore, Sarah and Lorraine Thorne. "Wild(er)ness: Reconfiguring the Geographies of Wildlife." *Transactions of the Institute of British Geographers* NS 23 (1998): 435–454.

Wheeler, Sara. "Love in a Cold Climate." *The Age* "Good Weekend," 17 December 2005.

Wienecke, Barbara. "Hard Times for Ross Sea Penguins." *Australian Antarctic Magazine* 3 (Autumn 2002): 37.

Wilbert, Chris. "Anti-This—Against-That: Resistances along a Human-Nonhuman Axis." In *Entanglements of Power: Geographies of Domination and Resistance*, edited by R. Paddison, C. Philo, P. Routledge, and J. Sharp, 238–55. London: Routledge, 2000.

Wilks, Samuel. "Vivisection: Its Pains and its Uses—III." *The Nineteenth Century* 10 (1881): 936–48.

Williams, Linda. "The Turn to the Animal in Contemporary Art—Why Now?" In *Unsettled Boundaries*, 6–7.

Williams, Murray, and Natasha Prince. "Tony Slaughter Critics Lashed: Ministry Backs Killing as SPCA Investigates." *Cape Argus*, 23 January 2007.

Williams, Raymond. *Marxism and Literature*. Oxford: Oxford University Press, 1977.

Wilson, Alexander. The Culture of Nature: North American Landscape from Disney to the Exxon Valdez. Cambridge, MA: Blackwell, 1992.

Wilson, Kelly. "Catching the Unique Rabbit: Why Pets Should Be Reclassified as Inimitable Property under the Law." *Cleveland State Law Review* 57 (2009): 167–96.

Wolch, Jennifer, and Jody Emel. "Witnessing the Animal Moment." In *Animal Geographies: Place, Politics and Identity in the Nature-Culture Borderlands*, edited by Jennifer Wolch and Jody Emel, 1–24. London: Verso, 1998.

Wolfe, Cary. *Animal Rites: American Culture, the Discourse of Species and Posthumanist Culture*. Chicago: University of Chicago Press, 2003.

———. "From *Dead Meat* to Glow in the Dark Bunnies: Seeing 'The Animal Question' in Contemporary Art." *Parallax* 38 (2006): 95–109.

———. "Introduction." In *Zoontontologies: The Question of the Animal*, edited by Cary Wolfe, ix–xxiii. Minneapolis: University of Minnesota Press, 2003.

———. "Old Orders for New: Ecology, Animal Rights and the Poverty of Humanism." *Diacritics* 28, no. 2 (1998): 21–40.

Yeld, John. "David, Cape Town's 'Gentleman Baboon', Killed by Four Dogs on Boland Farm." *Cape Argus*, 11 June 2007.

———. "Fears for Baboons as Two More Killed." *Cape Argus*, 24 May 2006.

Ziffer, Daniel. "Penguins Earn $55 Million in Cold, Hard, Cash." *The Age*, 21 November 2006.

Zogby International. "Americans: Make Disaster Plans for Pets, Too!" 14 October 2005. http://www.zogby.com/news/readnews.cfm?ID=1029 (accessed November 5, 2009).

Index